2001 | Higher

[BLANK]

Higher

Physics

2001 Exam

2002 Exam

2003 Exam

2004 Exam

2005 Exam

Leckie ✕ Leckie

First exam published in 2001.

Published by Leckie & Leckie, 8 Whitehill Terrace, St. Andrews, Scotland KY16 8RN tel: 01334 475656 fax: 01334 477392 enquiries@leckieandleckie.co.uk www.leckieandleckie.co.uk

ISBN 1-84372-350-6

A CIP Catalogue record for this book is available from the British Library.

Printed in Scotland by Scotprint.

Leckie & Leckie is a division of Granada Learning Limited, part of ITV plc.

Acknowledgements

Leckie & Leckie is grateful to the copyright holders, as credited at the back of the book, for permission to use their material.

Every effort has been made to trace the copyright holders and to obtain their permission for the use of copyright material.

Leckie & Leckie will gladly receive information enabling them to rectify any error or omission in subsequent editions.

X069/301

NATIONAL
QUALIFICATIONS
2001

MONDAY, 4 JUNE
9.00 AM – 11.30 AM

PHYSICS
HIGHER

Read Carefully

1   All questions should be attempted.

Section A (questions 1 to 20)

2   Check that the answer sheet is for Physics Higher (Section A).

3   Answer the questions numbered 1 to 20 on the answer sheet provided.

4   Fill in the details required on the answer sheet.

5   Rough working, if required, should be done only on this question paper, or on the first two pages of the answer book provided—**not** on the answer sheet.

6   For each of the questions 1 to 20 there is only **one** correct answer and each is worth 1 mark.

7   Instructions as to how to record your answers to questions 1–20 are given on page three.

Section B (questions 21 to 29)

8   Answer questions numbered 21 to 29 in the answer book provided.

9   Fill in the details on the front of the answer book.

10   Enter the question number clearly in the margin of the answer book beside each of your answers to questions 21 to 29.

11   Care should be taken to give an appropriate number of significant figures in the final answers to calculations.

SCOTTISH
QUALIFICATIONS
AUTHORITY

DATA SHEET
COMMON PHYSICAL QUANTITIES

Quantity	Symbol	Value	Quantity	Symbol	Value
Speed of light in vacuum	c	$3.00 \times 10^8 \,\mathrm{m\,s^{-1}}$	Mass of electron	m_e	9.11×10^{-31} kg
Magnitude of the charge on an electron	e	1.60×10^{-19} C	Mass of neutron	m_n	1.675×10^{-27} kg
Gravitational acceleration	g	$9.8 \,\mathrm{m\,s^{-2}}$	Mass of proton	m_p	1.673×10^{-27} kg
Planck's constant	h	6.63×10^{-34} J s			

REFRACTIVE INDICES

The refractive indices refer to sodium light of wavelength 589 nm and to substances at a temperature of 273 K.

Substance	Refractive index	Substance	Refractive index
Diamond	2·42	Water	1·33
Crown glass	1·50	Air	1·00

SPECTRAL LINES

Element	Wavelength/nm	Colour	Element	Wavelength/nm	Colour
Hydrogen	656	Red	Cadmium	644	Red
	486	Blue-green		509	Green
	434	Blue-violet		480	Blue
	410	Violet			
	397	Ultraviolet			
	389	Ultraviolet			
Sodium	589	Yellow			

	Lasers	
Element	Wavelength/nm	Colour
Carbon dioxide	9550 ⎱ 10590 ⎰	Infrared
Helium-neon	633	Red

PROPERTIES OF SELECTED MATERIALS

Substance	Density/ $\mathrm{kg\,m^{-3}}$	Melting Point/ K	Boiling Point/ K
Aluminium	2.70×10^3	933	2623
Copper	8.96×10^3	1357	2853
Ice	9.20×10^2	273	. . . .
Sea Water	1.02×10^3	264	377
Water	1.00×10^3	273	373
Air	1·29	. . . .	. . . .
Hydrogen	9.0×10^{-2}	14	20

The gas densities refer to a temperature of 273 K and a pressure of 1.01×10^5 Pa.

SECTION A

For questions 1 to 20 in this section of the paper, an answer is recorded on the answer sheet by indicating the choice A, B, C, D or E by a stroke made in ink in the appropriate box of the answer sheet—see the example below.

EXAMPLE

The energy unit measured by the electricity meter in your home is the

    A   ampere

    B   kilowatt-hour

    C   watt

    D   coulomb

    E   volt.

The correct answer to the question is B—kilowatt-hour. Record your answer by drawing a heavy vertical line joining the two dots in the appropriate box on your answer sheet in the column of boxes headed B. The entry on your answer sheet would now look like this:

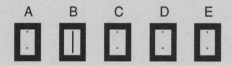

If after you have recorded your answer you decide that you have made an error and wish to make a change, you should cancel the original answer and put a vertical stroke in the box you now consider to be correct. Thus, if you want to change an answer D to an answer B, your answer sheet would look like this:

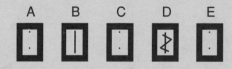

If you want to change back to an answer which has already been scored out, you should enter a tick (✓) to the RIGHT of the box of your choice, thus:

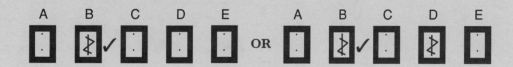

SECTION A

Answer questions 1–20 on the answer sheet.

1. Which one of the following pairs contains one vector quantity and one scalar quantity?

   A   Force, kinetic energy

   B   Power, speed

   C   Displacement, acceleration

   D   Work, potential energy

   E   Momentum, velocity

2. The diagram below shows the resultant of two vectors.

   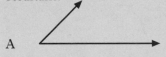

   Which of the diagrams below shows the vectors which could produce the above resultant?

   A

   B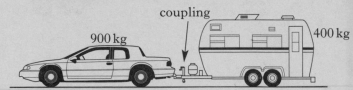

   C

   D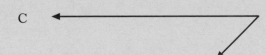

   E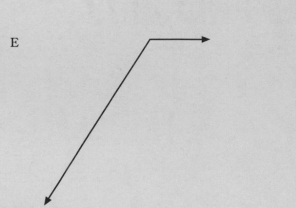

3. A helicopter is **descending** vertically at a constant speed of $3 \cdot 0\,\text{m s}^{-1}$. A sandbag is released from the helicopter. The sandbag hits the ground $5 \cdot 0\,\text{s}$ later.

   What was the height of the helicopter above the ground at the time the sandbag was released?

   A $15 \cdot 0\,\text{m}$

   B $49 \cdot 0\,\text{m}$

   C $107 \cdot 5\,\text{m}$

   D $122 \cdot 5\,\text{m}$

   E $137 \cdot 5\,\text{m}$

4. A car of mass $900\,\text{kg}$ pulls a caravan of mass $400\,\text{kg}$ along a straight, horizontal road with an acceleration of $2 \cdot 0\,\text{m s}^{-2}$.

   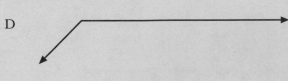

   Assuming that the frictional forces on the caravan are negligible, the tension in the coupling between the car and the caravan is

   A $400\,\text{N}$

   B $500\,\text{N}$

   C $800\,\text{N}$

   D $1800\,\text{N}$

   E $2600\,\text{N}$.

5. A rocket of mass $5 \cdot 0\,\text{kg}$ is travelling horizontally with a speed of $200\,\text{m s}^{-1}$ when it explodes into two parts. One part of mass $3 \cdot 0\,\text{kg}$ continues in the original direction with a speed of $100\,\text{m s}^{-1}$. The other part also continues in this same direction. Its speed is

   A $150\,\text{m s}^{-1}$

   B $200\,\text{m s}^{-1}$

   C $300\,\text{m s}^{-1}$

   D $350\,\text{m s}^{-1}$

   E $700\,\text{m s}^{-1}$.

6. A block floats in water and two other liquids X and Y at the levels shown.

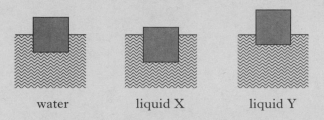

water   liquid X   liquid Y

Which of the following statements is/are correct?

  I The density of the material of the block is less than the density of water.

 II The density of liquid X is less than the density of water.

III The density of liquid X is greater than the density of liquid Y.

A   I only

B   II only

C   I and II only

D   I and III only

E   II and III only

7. Ice at −10 °C is heated until it becomes water at 80 °C.

The temperature change on the kelvin scale is

A   70 K

B   90 K

C   343 K

D   363 K

E   636 K.

8. In the diagrams below, each resistor has a resistance of 1·0 ohm.

Select the combination which has the **least** value of effective resistance between the terminals X and Y.

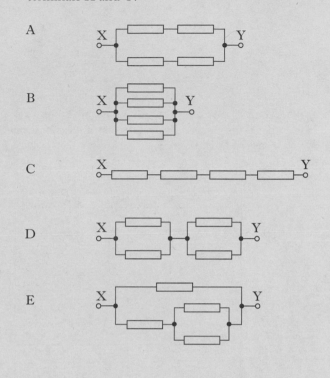

9. In the following circuit, the supply has negligible internal resistance.

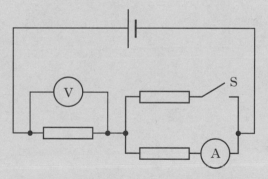

Switch S is now closed.

Which row in the table shows the effect on the ammeter and voltmeter readings?

	Ammeter reading	Voltmeter reading
A	increases	increases
B	increases	decreases
C	decreases	decreases
D	decreases	increases
E	decreases	remains the same

10. A supply with a sinusoidally alternating output of 6·0 V r.m.s. is connected to a 3·0 Ω resistor.

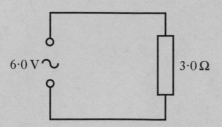

Which row in the following table shows the peak voltage across the resistor and the peak current in the circuit?

	Peak voltage/V	Peak current/A
A	$6\sqrt{2}$	$2\sqrt{2}$
B	$6\sqrt{2}$	2
C	6	2
D	$6\sqrt{2}$	$\dfrac{1}{2\sqrt{2}}$
E	6	$2\sqrt{2}$

11. A resistor and an ammeter are connected to a signal generator having an output of constant amplitude and variable frequency.

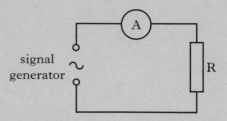

Which of the following graphs shows the correct relationship between the current I in the resistor and the output frequency f of the signal generator?

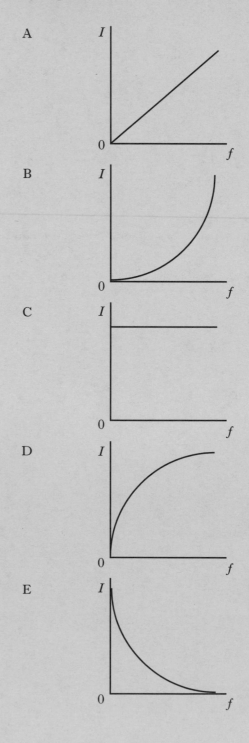

12. Which of the following statements is/are true for an ideal op-amp?

   I It has infinite input resistance.

   II Both input pins are at the same potential.

   III The input current to the op-amp is zero.

  A I only

  B II only

  C I and II only

  D II and III only

  E I, II and III

13. An op-amp circuit is shown in the diagram.

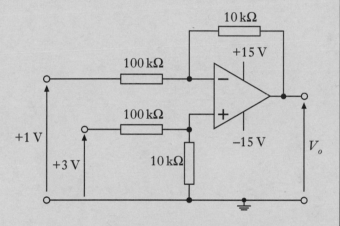

What is the output voltage V_o?

  A −20 V

  B −2 V

  C −0·2 V

  D 0·2 V

  E 20 V

14. The energy of a water wave depends on its

  A speed

  B wavelength

  C frequency

  D period

  E amplitude.

15. S_1 and S_2 are sources of coherent waves which produce an interference pattern along the line XY.

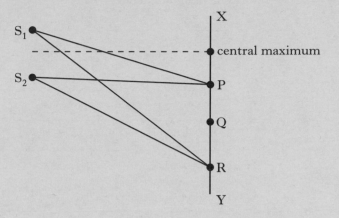

The first maximum occurs at P, where $S_1P = 20$ cm and $S_2P = 18$ cm.

For the third maximum, at R, the path difference $(S_1R - S_2R)$ is

  A 3 cm

  B 4 cm

  C 5 cm

  D 6 cm

  E 8 cm.

16. The spectrum of white light from a filament lamp may be viewed using a prism or a grating.

A student, asked to compare the spectra formed by the two methods, made the following statements.

  I The prism produces a spectrum by refraction. The grating produces a spectrum by interference.

  II The spectrum formed by the prism shows all the wavelengths present in the white light. The spectrum formed by the grating shows only a few specific wavelengths.

  III The prism produces a single spectrum. The grating produces more than one spectrum.

Which of the above statements is/are true?

  A I only

  B II only

  C I and II only

  D I and III only

  E I, II and III

17. Red light passes from air into water.

What happens to the speed and frequency of the light when it enters the water?

	Speed	Frequency
A	increases	increases
B	increases	stays constant
C	decreases	stays constant
D	decreases	decreases
E	stays constant	decreases

18. The intensity of light from a point source is $20\,\mathrm{W\,m^{-2}}$ at a distance of $5 \cdot 0\,\mathrm{m}$ from the source.

What is the intensity of the light at a distance of $25\,\mathrm{m}$ from the source?

A $0 \cdot 032\,\mathrm{W\,m^{-2}}$

B $0 \cdot 80\,\mathrm{W\,m^{-2}}$

C $1 \cdot 2\,\mathrm{W\,m^{-2}}$

D $4 \cdot 0\,\mathrm{W\,m^{-2}}$

E $100\,\mathrm{W\,m^{-2}}$

19. Ultraviolet radiation causes the emission of photoelectrons from a zinc plate.

The intensity of the ultraviolet radiation is increased. Which row in the following table shows the effect of this change?

	Maximum kinetic energy of a photoelectron	Number of photoelectrons per second
A	increases	no change
B	no change	increases
C	no change	no change
D	increases	increases
E	decreases	increases

20. Under certain conditions, a nucleus of nitrogen absorbs an alpha particle to form the nucleus of another element and releases a single particle.

Which one of the following statements correctly describes this process?

A $^{14}_{7}\mathrm{N} + {}^{3}_{2}\mathrm{He} \rightarrow {}^{16}_{9}\mathrm{F} + {}^{1}_{0}\mathrm{n}$

B $^{14}_{7}\mathrm{N} + {}^{4}_{2}\mathrm{He} \rightarrow {}^{17}_{10}\mathrm{N} + {}^{0}_{-1}\mathrm{e}$

C $^{14}_{7}\mathrm{N} + {}^{3}_{2}\mathrm{He} \rightarrow {}^{16}_{8}\mathrm{O} + {}^{1}_{1}\mathrm{p}$

D $^{14}_{7}\mathrm{N} + {}^{4}_{2}\mathrm{He} \rightarrow {}^{18}_{9}\mathrm{F} + 2\,{}^{0}_{-1}\mathrm{e}$

E $^{14}_{7}\mathrm{N} + {}^{4}_{2}\mathrm{He} \rightarrow {}^{17}_{8}\mathrm{O} + {}^{1}_{1}\mathrm{p}$

SECTION B

Write your answers to questions 21 to 29 in the answer book.   *Marks*

21. (*a*) A box of mass 18 kg is at rest on a horizontal frictionless surface.
A force of 4·0 N is applied to the box at an angle of 26° to the horizontal.

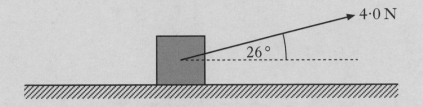

(i)   Show that the horizontal component of this force is 3·6 N.

(ii) Calculate the acceleration of the box along the horizontal surface.

(iii) Calculate the horizontal distance travelled by the box in a time of
7·0 s.   5

(*b*) The box is replaced at rest at its starting position.

The force of 4·0 N is now applied to the box at an angle of less than 26° to
the horizontal.

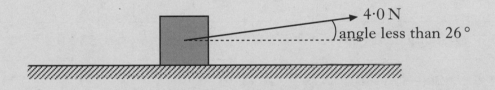

The force is applied for a time of 7·0 s as before.

How does the distance travelled by the box compare with your answer to
part (*a*)(iii)?

You must justify your answer.   2

(7)

[**Turn over**

Marks

22. (a) In an experiment to find the density of air, a student first measures the mass of a flask full of air as shown below.

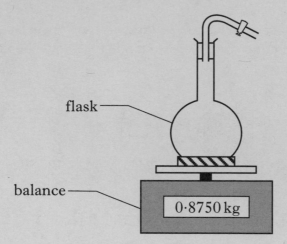

The air is now removed from the flask and the mass of the evacuated flask measured.

This procedure is repeated a number of times and the following table of measurements is obtained.

	Experiment number					
	1	2	3	4	5	6
Mass of flask and air/kg	0·8750	0·8762	0·8748	0·8755	0·8760	0·8757
Mass of evacuated flask/kg	0·8722	0·8736	0·8721	0·8728	0·8738	0·8732
Mass of air removed/kg						

The volume of the flask is measured as $2 \cdot 0 \times 10^{-3} \, \text{m}^3$.

(i) Copy and complete the **bottom row** of the table.

(ii) Calculate the mean mass of air removed from the flask **and** the random uncertainty in this mean. Express the mean mass and the random uncertainty in kilograms.

(iii) Use these measurements to calculate the density of air.

(iv) Another student carries out the same experiment using a flask of larger volume.

Explain why this is a better design for the experiment.

6

Marks

22. (continued)

(b) The cylinder of a bicycle pump has a length of 360 mm as shown in the diagram.

The outlet of the pump is sealed.

The piston is pushed inwards until it is 160 mm from the outlet.

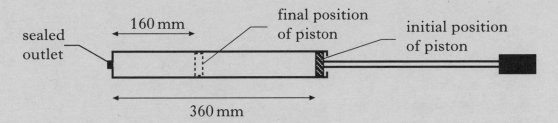

The initial pressure of the air in the pump is $1 \cdot 0 \times 10^5$ Pa.

(i) Assuming that the temperature of the air trapped in the cylinder remains constant, calculate the final pressure of the trapped air.

(ii) State one other assumption you have made for this calculation.

(iii) Use the kinetic model to explain what happens to the pressure of the trapped air as its volume decreases.

5

(11)

[Turn over

Marks

23. Beads of liquid moving at high speed are used to move threads in modern weaving machines.

   (*a*) In one design of machine, beads of water are accelerated by jets of air as shown in the diagram.

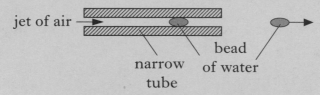

   jet of air

   narrow tube

   bead of water

   Each bead has a mass of 2.5×10^{-5} kg.

   When designing the machine, it was estimated that each bead of water would start from rest and experience a constant unbalanced force of 0.5 N for a time of 3.0 ms.

   (i) Calculate:

   (A)   the impulse on a bead of water;

   (B)   the speed of the bead as it emerges from the tube.

   (ii) In practice the force on a bead varies.

   The following graph shows how the actual unbalanced force exerted on each bead of water varies with time.

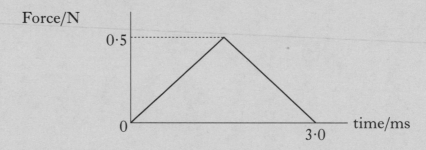

   Force/N

   0·5

   0

   3·0

   time/ms

   Use information from this graph to show that the bead leaves the tube with a speed equal to half of the value calculated in part (i)(B).   **6**

   (*b*) Another design of machine uses beads of oil and two metal plates X and Y.

   The potential difference between these plates is 5.0×10^{3} V.

   Each bead of oil has a mass of 4.0×10^{-5} kg and is given a negative charge of 6.5×10^{-6} C.

   The bead accelerates from rest at plate X and passes through a hole in plate Y.

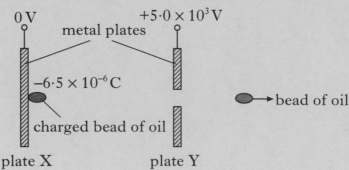

   0 V

   $+5.0 \times 10^{3}$ V

   metal plates

   -6.5×10^{-6} C

   bead of oil

   charged bead of oil

   plate X

   plate Y

   Neglecting air friction, calculate the speed of the bead at plate Y.   **3**

Marks

24. (*a*) The following circuit is used to measure the e.m.f. and the internal resistance of a battery.

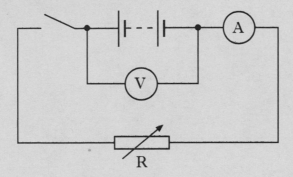

Readings of current and potential difference from this circuit are used to produce the following graph.

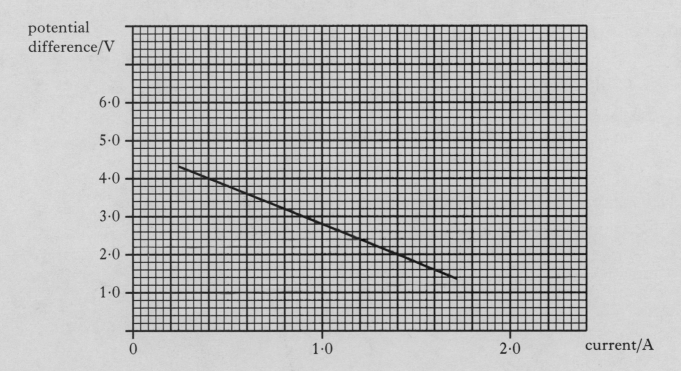

Use information from the graph to find:

(i) the e.m.f. of the battery, in volts;

(ii) the internal resistance of the battery.   **3**

(*b*) A car battery has an e.m.f. of 12 V and an internal resistance of $0.050\,\Omega$.

(i) Calculate the short circuit current for this battery.

(ii) The battery is now connected in series with a lamp. The resistance of the lamp is $2.5\,\Omega$. Calculate the power dissipated in the lamp.   **5**

           (8)

[Turn over

Marks

25. (*a*) The following diagram shows a circuit that is used to investigate the charging of a capacitor.

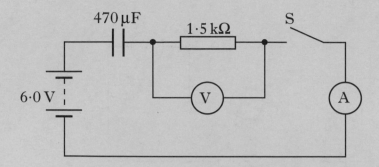

The capacitor is initially uncharged.

The capacitor has a capacitance of 470 µF and the resistor has a resistance of 1·5 kΩ.

The battery has an e.m.f. of 6·0 V and negligible internal resistance.

(i) Switch S is now closed. What is the initial current in the circuit?

(ii) How much energy is stored in the capacitor when it is fully charged?

(iii) What change could be made to this circuit to ensure that the **same** capacitor stores **more** energy?   **5**

(*b*) A capacitor is used to provide the energy for an electronic flash in a camera.

When the flash is fired, $6·35 \times 10^{-3}$ J of the stored energy is emitted as light.

The mean value of the frequency of photons of light from the flash is $5·80 \times 10^{14}$ Hz.

Calculate the number of photons emitted in each flash of light.   **3**

(8)

Marks

26. (*a*) An op-amp is connected in a circuit as shown below.

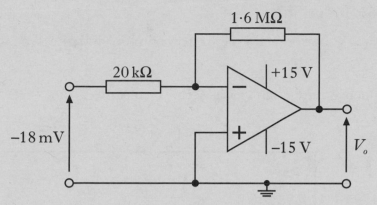

(i) In which mode is the op-amp operating?

(ii) A voltage of −18 mV is connected to the input. Calculate the output voltage V_o.

(iii) The supply voltage is now reduced from ±15 V to ±12 V.

State any effect this change has on the output voltage. You must justify your answer. **4**

(*b*) A student connects an op-amp as shown in the following diagram. An alternating voltage of peak value 5·0 V is connected to the input as shown.

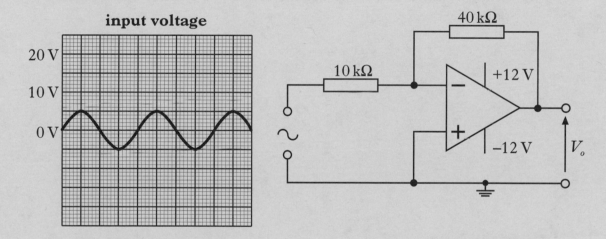

input voltage

The sketch below shows the student's attempt to draw the corresponding output voltage.

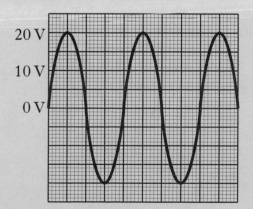

State the **two** mistakes in the student's sketch. **2**

(6)

Marks

27. (a) Light of wavelength 486×10^{-9} m is viewed using a grating with a slit spacing of $2 \cdot 16 \times 10^{-6}$ m.

Calculate the angle between the central maximum and the second order maximum.

2

(b) A ray of monochromatic light passes from air into a block of glass as shown.

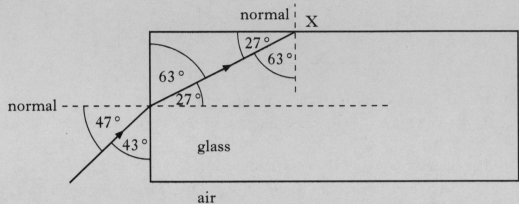

(i) Using information from the diagram, show that the refractive index of the glass for this light is 1·61.

(ii) Show by calculation whether the ray is totally internally reflected at point X.

4

(6)

Marks

28. (*a*) In a laser, the light is produced by stimulated emission of radiation.

Explain the term "stimulated emission" by making reference to the energy levels in atoms. **2**

(*b*) A laser beam is shone on to a screen which is marked with a grid.

The beam produces a uniformly lit spot of radius 5.00×10^{-4} m as shown.

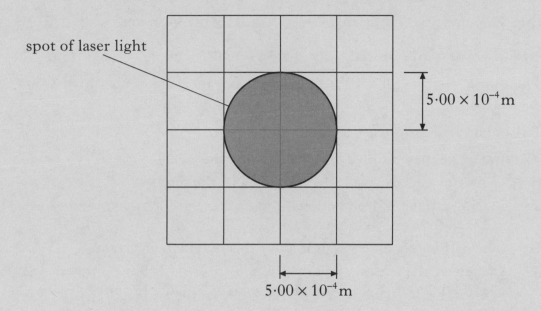

spot of laser light

5.00×10^{-4} m

5.00×10^{-4} m

    (i) The intensity of the spot of light on the screen is $1020\,\text{W m}^{-2}$.

    Calculate the power of the laser beam.

    (ii) The distance between the screen and the laser is now doubled.

    State how the radius of the spot now compares with the one shown in the diagram.

    You must justify your answer. **5**

(7)

[Turn over

Marks

29. (*a*) The following statement represents a nuclear reaction.

$$^{239}_{94}\text{Pu} + {}^{1}_{0}\text{n} \longrightarrow {}^{137}_{52}\text{Te} + {}^{100}_{42}\text{Mo} + 3{}^{1}_{0}\text{n} + \text{energy}$$

The total mass of the particles before the reaction is $3{\cdot}9842 \times 10^{-25}$kg and the total mass of the particles after the reaction is $3{\cdot}9825 \times 10^{-25}$kg.

  (i)   State and explain whether this reaction is spontaneous or induced.

 (ii)   Calculate the energy, in joules, released by this reaction.   **3**

(*b*) A radioactive source is used to irradiate a sample of tissue of mass $0{\cdot}50$ kg.

The tissue absorbs $9{\cdot}6 \times 10^{-5}$ J of energy from the radiation emitted from the source.

The radiation has a quality factor of 1.

  (i)   Calculate the absorbed dose received by the tissue.

 (ii)   Calculate the dose equivalent received by the tissue.

(iii)   Placing a sheet of lead between the source and the tissue would have reduced the dose received by the tissue.

       The half-value thickness of lead for this radiation is 40 mm.

       Calculate the thickness of lead which would have limited the absorbed dose to one eighth of the value calculated in part (*b*)(i).   **5**

                                                                 (8)

[END OF QUESTION PAPER]

[BLANK]

X069/301

NATIONAL
QUALIFICATIONS
2002

WEDNESDAY, 22 MAY
1.00 PM – 3.30 PM

PHYSICS
HIGHER

Read Carefully

1   All questions should be attempted.

Section A (questions 1 to 20)

2   Check that the answer sheet is for Physics Higher (Section A).

3   Answer the questions numbered 1 to 20 on the answer sheet provided.

4   Fill in the details required on the answer sheet.

5   Rough working, if required, should be done only on this question paper, or on the first two pages of the answer book provided—**not** on the answer sheet.

6   For each of the questions 1 to 20 there is only **one** correct answer and each is worth 1 mark.

7   Instructions as to how to record your answers to questions 1–20 are given on page three.

Section B (questions 21 to 30)

8   Answer questions numbered 21 to 30 in the answer book provided.

9   Fill in the details on the front of the answer book.

10   Enter the question number clearly in the margin of the answer book beside each of your answers to questions 21 to 30.

11   Care should be taken to give an appropriate number of significant figures in the final answers to calculations.

DATA SHEET
COMMON PHYSICAL QUANTITIES

Quantity	Symbol	Value	Quantity	Symbol	Value
Speed of light in vacuum	c	$3 \cdot 00 \times 10^8$ m s^{-1}	Mass of electron	m_e	$9 \cdot 11 \times 10^{-31}$ kg
Magnitude of the charge on an electron	e	$1 \cdot 60 \times 10^{-19}$ C	Mass of neutron	m_n	$1 \cdot 675 \times 10^{-27}$ kg
Gravitational acceleration	g	$9 \cdot 8$ m s^{-2}	Mass of proton	m_p	$1 \cdot 673 \times 10^{-27}$ kg
Planck's constant	h	$6 \cdot 63 \times 10^{-34}$ J s			

REFRACTIVE INDICES
The refractive indices refer to sodium light of wavelength 589 nm and to substances at a temperature of 273 K.

Substance	Refractive index	Substance	Refractive index
Diamond	2·42	Water	1·33
Crown glass	1·50	Air	1·00

SPECTRAL LINES

Element	Wavelength/nm	Colour	Element	Wavelength/nm	Colour
Hydrogen	656	Red	Cadmium	644	Red
	486	Blue-green		509	Green
	434	Blue-violet		480	Blue
	410	Violet			
	397	Ultraviolet			
	389	Ultraviolet			
Sodium	589	Yellow			

Lasers		
Element	Wavelength/nm	Colour
Carbon dioxide	9550 } 10590 }	Infrared
Helium-neon	633	Red

PROPERTIES OF SELECTED MATERIALS

Substance	Density/ kg m^{-3}	Melting Point/ K	Boiling Point/ K
Aluminium	$2 \cdot 70 \times 10^3$	933	2623
Copper	$8 \cdot 96 \times 10^3$	1357	2853
Ice	$9 \cdot 20 \times 10^2$	273	. . . .
Sea Water	$1 \cdot 02 \times 10^3$	264	377
Water	$1 \cdot 00 \times 10^3$	273	373
Air	$1 \cdot 29$	. . . .	. . . .
Hydrogen	$9 \cdot 0 \times 10^{-2}$	14	20

The gas densities refer to a temperature of 273 K and a pressure of $1 \cdot 01 \times 10^5$ Pa.

SECTION A

For questions 1 to 20 in this section of the paper, an answer is recorded on the answer sheet by indicating the choice A, B, C, D or E by a stroke made in ink in the appropriate box of the answer sheet—see the example below.

EXAMPLE

The energy unit measured by the electricity meter in your home is the

    A   ampere

    B   kilowatt-hour

    C   watt

    D   coulomb

    E   volt.

The correct answer to the question is B—kilowatt-hour. Record your answer by drawing a heavy vertical line joining the two dots in the appropriate box on your answer sheet in the column of boxes headed B. The entry on your answer sheet would now look like this:

If after you have recorded your answer you decide that you have made an error and wish to make a change, you should cancel the original answer and put a vertical stroke in the box you now consider to be correct. Thus, if you want to change an answer D to an answer B, your answer sheet would look like this:

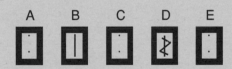

If you want to change back to an answer which has already been scored out, you should enter a tick (✓) to the RIGHT of the box of your choice, thus:

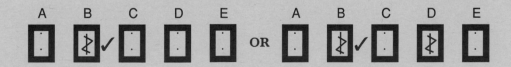

SECTION A

Answer questions 1–20 on the answer sheet.

1. The following graph shows how the displacement of an object varies with time.

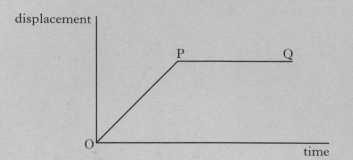

Which row of the table below best describes the motion of this object?

	From O to P	*From P to Q*
A	constant acceleration	constant velocity
B	zero velocity	constant deceleration
C	constant velocity	zero velocity
D	zero velocity	constant velocity
E	constant velocity	constant deceleration

2. Which of the following velocity-time graphs best describes a ball being thrown vertically into the air and returning to the thrower's hand?

A

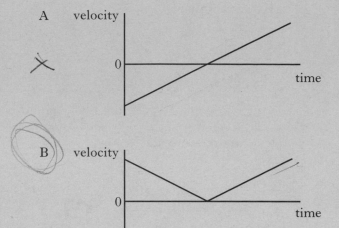

B

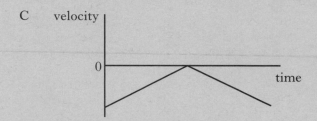

C

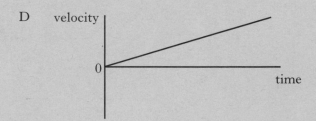

D

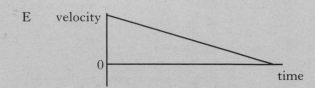

E velocity

3. A force of 180 N is applied vertically upwards to a box of mass 15 kg.

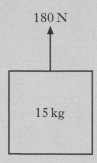

180 N

15 kg

Assuming that the acceleration due to gravity is $9 \cdot 8 \, \text{m s}^{-2}$, the acceleration of the box is

A $2 \cdot 2 \, \text{m s}^{-2}$

B $7 \cdot 6 \, \text{m s}^{-2}$

C $9 \cdot 8 \, \text{m s}^{-2}$

D $12 \cdot 0 \, \text{m s}^{-2}$

E $19 \cdot 6 \, \text{m s}^{-2}$.

4. A box of mass 10 kg rests on an inclined plane. The component of the weight of the box acting down the incline is 50 N. A force of 300 N, parallel to the plane, is applied to the box as shown.

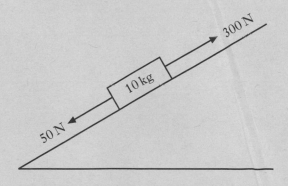

300 N

10 kg

50 N

The box accelerates at $10 \, \text{m s}^{-2}$ up the plane.

The size of the force of friction opposing the motion of the box is

A   50 N

B   100 N

C   150 N

D   200 N

E   250 N.

5. A flat bottomed test-tube containing aluminium rivets is floated in liquid A.

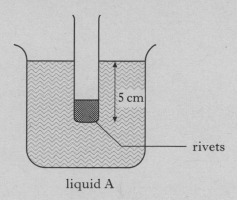

5 cm

rivets

liquid A

The bottom of the test-tube is at a depth of 5 cm below the surface.

The same test-tube and aluminium rivets are then floated in liquid B.

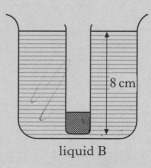

8 cm

liquid B

The bottom of the test-tube is at a depth of 8 cm below the surface.

Which of the following statement(s) is/are true?

I   In each liquid the pressure at the bottom of the test-tube is the same.

II   The density of liquid A is greater than the density of liquid B.

III   In each liquid the upthrust on the bottom of the test-tube is the same.

A   I only

B   II only

C   I and II only

D   II and III only

E   I, II and III

[Turn over

6. A helium filled balloon of mass $1\cdot5\,kg$ floats at a constant height of $100\,m$. The acceleration due to gravity is $9\cdot8\,m\,s^{-2}$.

The upthrust on the balloon is

   A $0\,N$ ✓

   B $1\cdot5\,N$

   Ⓒ $14\cdot7\,N$

   D $150\,N$

   E $1470\,N$.

7. A sealed hollow buoy drifts from warm Atlantic waters into colder Arctic waters.

The volume of the buoy remains constant.

The pressure of the air trapped inside the buoy changes.

This is because the pressure of the trapped air is

   Ⓐ   directly proportional to the kelvin temperature

   B   inversely proportional to the kelvin temperature ✓

   C   inversely proportional to the volume of the air in the buoy

   D   inversely proportional to the celsius temperature

   E   directly proportional to the celsius temperature.

8. In the following circuit the current from the battery is $3\,A$.

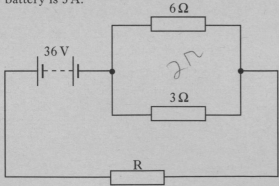

Assuming that the battery has negligible internal resistance, the resistance of resistor R is

   A $3\,\Omega$

   B $4\,\Omega$

   Ⓒ $10\,\Omega$

   D $12\,\Omega$

   E $18\,\Omega$.

9. The diagram below shows a balanced Wheatstone bridge where all the resistors have different values.

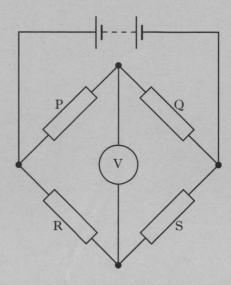

Which change(s) would make the bridge unbalanced?

   I   Interchange resistors P and S.

   II Interchange resistors P and Q.

   III Change the e.m.f. of the battery.

   A   I only

   Ⓑ   II only

   C   III only

   D   II and III only ✓

   E   I and III only

10. A student sets up the following circuit.

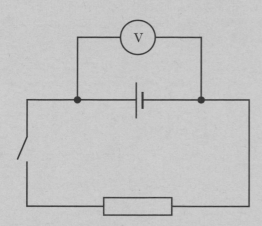

When the switch is open, the student notes that the reading on the voltmeter is $1\cdot5$ V. The switch is then closed and the reading falls to $1\cdot3$ V.

The decrease of $0\cdot2$ V is referred to as the

A   e.m.f.

B   lost volts

C   peak voltage

D   r.m.s. voltage

E   terminal potential difference.

11. The unit for capacitance can be written as

A $V C^{-1}$

B $C V^{-1}$

C $J s^{-1}$

D $C J^{-1}$

E $J C^{-1}$.

12. Which of the following statements about capacitors is/are true?

  I   Capacitors are used to block a.c. signals.

  II Capacitors are used to block d.c. signals.

  III Capacitors can store energy.

  IV Capacitors can store electric charge.

A   I only

B   I and III only

C   II and III only

D   II, III and IV only

E   III and IV only

13. The operational amplifier connected in the circuit below is powered by a supply of +15 V and −15 V.

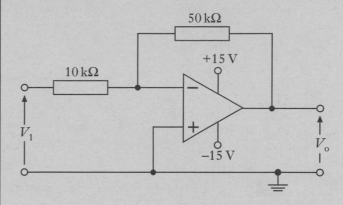

The input voltage V_1 is +5 V. The most likely value for the output voltage V_o is

A   −25 V

B   −13 V

C   −1 V

D   +13 V

E   +25 V.

14. The amplifier shown below has an output voltage of $5\cdot0$ V.

Input voltage V_1 is originally $0\cdot5$ V and input voltage V_2 is originally $0\cdot6$ V.

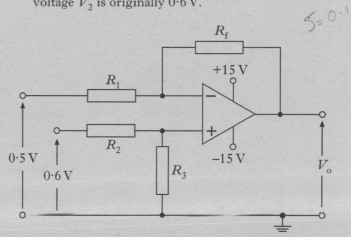

$R_1 = R_2$ and $R_f = R_3$

The input voltages V_1 and V_2 are increased to $1\cdot0$ V and $1\cdot2$ V respectively.

The output voltage V_o is now

A $0\cdot2$ V

B $2\cdot2$ V

C $5\cdot0$ V

D $6\cdot0$ V

E   10 V.

15. Microwave radiation is incident on a metal plate which has 2 slits, P and Q. A microwave receiver is moved from R to S, and detects a series of maxima and minima of intensity at the positions shown.

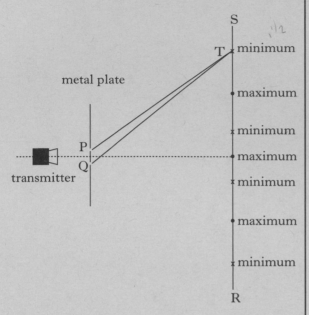

The microwave radiation has a wavelength of 4 cm.

The path difference between PT and QT is

A   2 cm

B   3 cm

C   4 cm

D   5 cm

(E)   6 cm.

16. Light of frequency $5 \cdot 0 \times 10^{14}$ Hz passes from air into a block of glass of refractive index $1 \cdot 5$.

Which row in the following table gives the correct values for the velocity, frequency and wavelength of the light in the glass?

	velocity/m s^{-1}	frequency/Hz	wavelength/m
A	$2 \cdot 0 \times 10^{8}$	$5 \cdot 0 \times 10^{14}$	$4 \cdot 0 \times 10^{-7}$
B	$3 \cdot 0 \times 10^{8}$	$5 \cdot 0 \times 10^{14}$	$6 \cdot 0 \times 10^{-7}$
C	$3 \cdot 0 \times 10^{8}$	$3 \cdot 3 \times 10^{14}$	$6 \cdot 0 \times 10^{-7}$
D	$2 \cdot 0 \times 10^{8}$	$3 \cdot 3 \times 10^{14}$	$6 \cdot 0 \times 10^{-7}$
E	$3 \cdot 0 \times 10^{8}$	$3 \cdot 3 \times 10^{14}$	$4 \cdot 0 \times 10^{-7}$

17. In a laser, a photon of radiation is emitted when an electron makes a transition from a higher energy level to a lower level, as shown below.

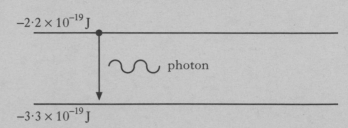

The energy in each pulse of radiation from the laser is 10 J. How many photons are there in each pulse?

A $1 \cdot 8 \times 10^{19}$

B $3 \cdot 0 \times 10^{19}$

C $3 \cdot 7 \times 10^{19}$

D $4 \cdot 5 \times 10^{19}$

(E) $9 \cdot 1 \times 10^{19}$

Turn over

18. In a darkened room, a small lamp is placed 2 cm from a photodiode which is connected in the circuit as shown. The lamp may be regarded as a point source. The reading on the ammeter is 27 µA.

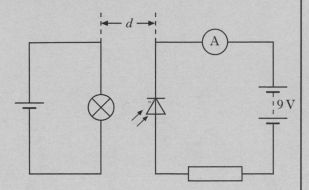

Which graph shows correctly how the ammeter reading changes as the distance d between the lamp and the photodiode is increased to 6 cm?

A
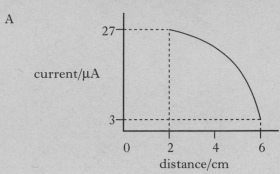

B
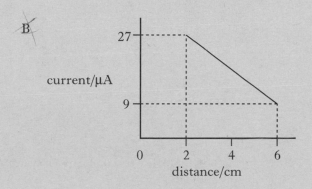

C
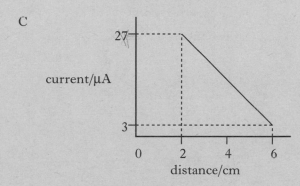

D
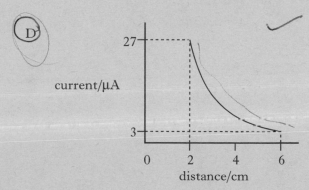

E
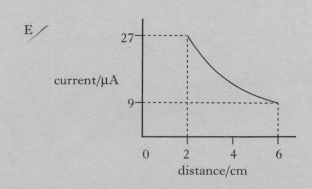

19. Which row of the table shows the correct values of x, y and z for the nuclear reaction described below?

$$^{214}_{x}\text{Pb} \rightarrow \,^{y}_{83}\text{Bi} + \,^{0}_{z}\text{e}$$

	x	y	z
A	84	214	1
B	83	210	4
C	85	214	2
D	82	214	−1
E	82	210	−1

20. The risk of biological harm from exposure to radiation depends on

 I the absorbed dose

 II the body organs exposed

 III the type of radiation.

Which statement(s) is/are true?

A I only

B II only

C III only

D II and III only

E I, II and III

[SECTION B begins on *Page twelve*]

SECTION B

Write your answers to questions 21 to 30 in the answer book.

21. A basketball is held below a motion sensor. The basketball is released from rest and falls onto a wooden block. The motion sensor is connected to a computer so that graphs of the motion of the basketball can be displayed.

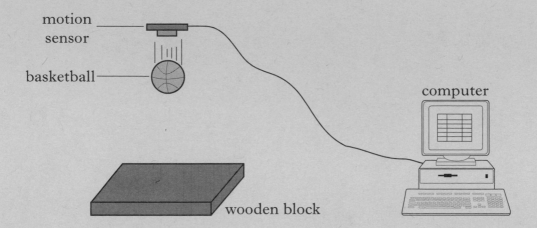

A displacement-time graph for the motion of the basketball from the instant of its release is shown.

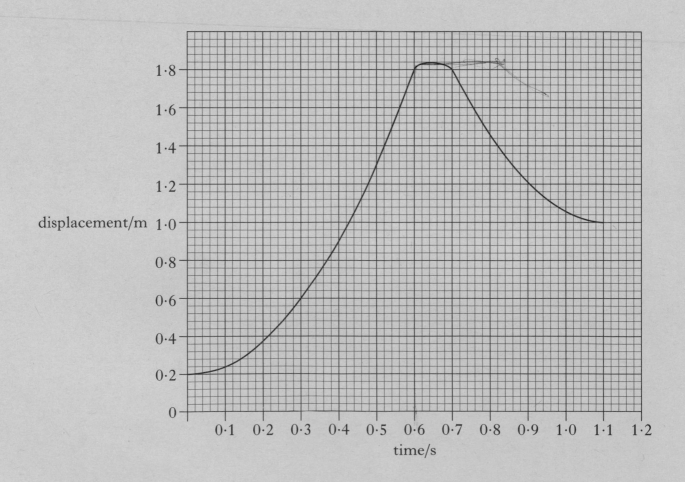

Marks

21. (continued)

(*a*) (i) What is the distance between the motion sensor and the top of the basketball when it is released?

    (ii) How far does the basketball fall before it hits the wooden block?

    (iii) Show, by calculation, that the acceleration of the basketball as it falls is $8 \cdot 9 \, \text{m s}^{-2}$.　　**3**

(*b*) The basketball is now dropped several times from the same height. The following values are obtained for the acceleration of the basketball.

$$8 \cdot 9 \, \text{m s}^{-2} \qquad 9 \cdot 1 \, \text{m s}^{-2} \qquad 8 \cdot 4 \, \text{m s}^{-2} \qquad 8 \cdot 5 \, \text{m s}^{-2} \qquad 9 \cdot 0 \, \text{m s}^{-2}$$

Calculate:

    (i) the mean of these values;

    (ii) the approximate random uncertainty in the mean.　　**3**

(*c*) The wooden block is replaced by a block of sponge of the same dimensions. The experiment is repeated and a new graph obtained.

Describe and explain any **two** differences between this graph and the original graph.　　**2**

(8)

[Turn over

Marks

22. A technician designs the apparatus shown in the diagram to investigate the relationship between the temperature and pressure of a fixed mass of nitrogen which is kept at a constant volume.

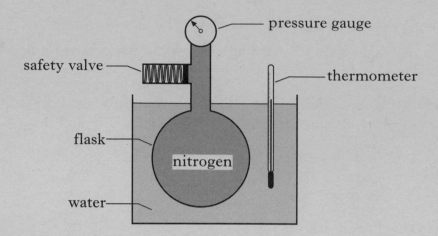

(a) The pressure of the nitrogen is 109 kPa when its temperature is 15 °C. The temperature of the nitrogen rises to 45 °C.

Calculate the new pressure of the nitrogen in the flask.

2

(b) Explain, in terms of the movement of gas molecules, what happens to the pressure of the nitrogen as its temperature is increased.

2

(c) The technician has fitted a safety valve to the apparatus.

A diagram of the valve is shown below.

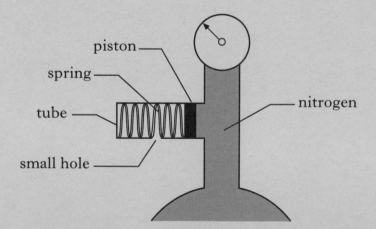

The piston of cross-sectional area $4.0 \times 10^{-6} \, \text{m}^2$ is attached to the spring. The piston is free to move along the tube.

The following graph shows how the length of the spring varies with the force exerted by the nitrogen on the piston.

Marks

22. **(c)** **(continued)**

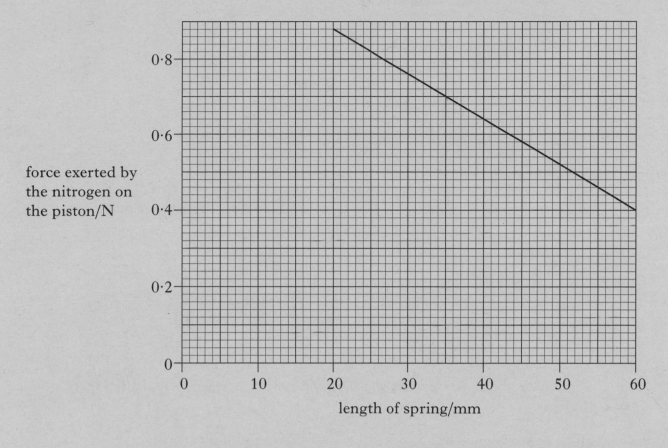

force exerted by
the nitrogen on
the piston/N

length of spring/mm

(i) Calculate the force exerted by the nitrogen on the piston when the reading on the pressure gauge is 1.75×10^5 Pa.

(ii) What is the length of the spring in the safety valve when the pressure of the nitrogen is 1.75×10^5 Pa? **3**

(d) The technician decides to redesign the apparatus so that the bulb of the thermometer is placed inside the flask.

Give **one** reason why this improves the design of the apparatus. **1**

(8)

[Turn over

Marks

23. (a) A space vehicle of mass 2500 kg is moving with a constant speed of 0·50 m s⁻¹ in the direction shown. It is about to dock with a space probe of mass 1500 kg which is moving with a constant speed in the opposite direction.

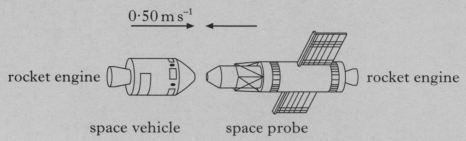

$0{\cdot}50\,\mathrm{m\,s^{-1}}$

rocket engine

space vehicle   space probe

rocket engine

After docking, the space vehicle and space probe move off together at 0·20 m s⁻¹ in the original direction in which the space vehicle was moving.

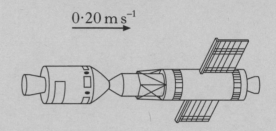

$0{\cdot}20\,\mathrm{m\,s^{-1}}$

Calculate the speed of the space probe before it docked with the space vehicle.

2

(b) The space vehicle has a rocket engine which produces a constant thrust of 1000 N. The space probe has a rocket engine which produces a constant thrust of 500 N.

The space vehicle and space probe are now brought to rest from their combined speed of 0·20 m s⁻¹.

  (i) Which rocket engine was switched on to bring the vehicle and probe to rest?

  (ii) Calculate the time for which this rocket engine was switched on. You may assume that a negligible mass of fuel was used during this time.

3

(c) The space vehicle and space probe are to be moved from their stationary position at A and brought to rest at position B, as shown.

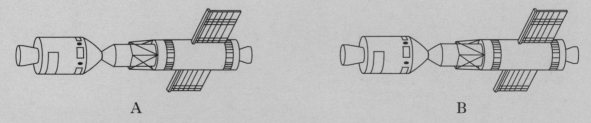

A

B

Explain clearly how the rocket engines of the space vehicle and the space probe are used to complete this manoeuvre.

Your explanation must include an indication of the relative time for which each rocket engine must be fired.

You may assume that a negligible mass of fuel is used during this manoeuvre.

2

Marks

24. A battery has an e.m.f. of 6·0 V and internal resistance of 2·0 Ω.

(a) What is meant by an *e.m.f. of 6·0 V*? 1

(b) The battery is connected in series with two resistors, R₁ and R₂. Resistor R₁ has a resistance of 20 Ω.

The reading on the ammeter is 200 mA.

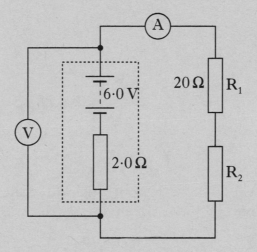

(i) Show by calculation that R₂ has a resistance of 8·0 Ω.

(ii) Calculate the reading on the voltmeter. 4

(c) The battery is now connected to two identical lamps as shown below.

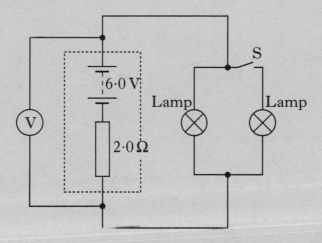

Describe and explain what happens to the reading on the voltmeter when switch S is closed. 2

(7)

[Turn over

Marks

25. (*a*) The circuit below is used to investigate the charging of a $2000\,\mu F$ capacitor. The d.c. supply has negligible internal resistance.

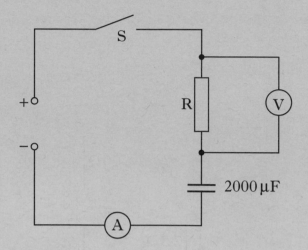

The graphs below show how the potential difference V_R across the **resistor** and the current I in the circuit vary with time from the instant switch S is closed.

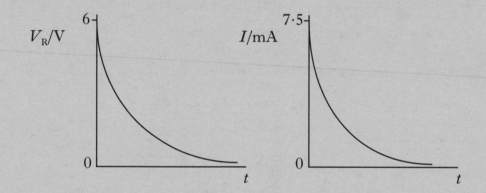

(i) What is the potential difference across the capacitor when it is fully charged?

(ii) Calculate the energy stored in the capacitor when it is fully charged.

(iii) Calculate the resistance of R in the circuit above.

5

Marks

25. **(continued)**

(b) The circuit below is used to investigate the charging and discharging of a capacitor.

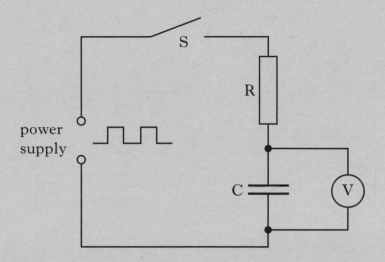

The graph below shows how the power supply voltage varies with time after switch S is closed.

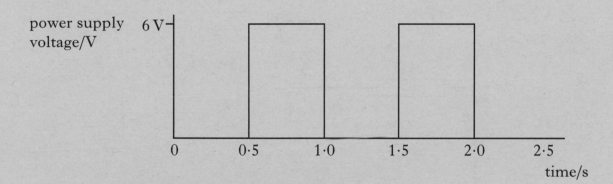

The capacitor is initially uncharged.

The capacitor charges fully in 0·3 s and discharges fully in 0·3 s.

Sketch a graph of the reading on the voltmeter for the first 2·5 s after switch S is closed.

The axes on your graph must have the same numerical values as those in the above graph.

2

(7)

[Turn over

Marks

26. An alternating voltage signal displayed on an oscilloscope screen is shown below. The peak voltage is 6·0 V and the time base setting is 2 ms/cm.

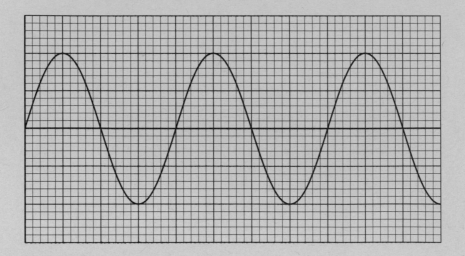

(a) Calculate the frequency of the signal. 2

(b) This alternating voltage is used as the input voltage V_1 for the operational amplifier circuit shown below. R_f is a variable resistor.

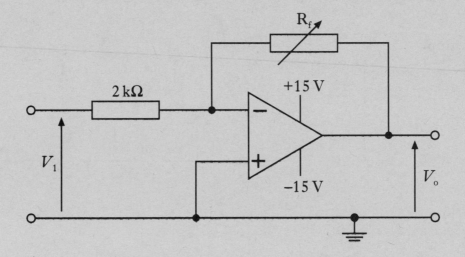

   (i) In what mode is the operational amplifier operating?

   (ii) The variable resistor R_f is set at 3·0 kΩ.

     (A) On square ruled paper, sketch a graph of the output voltage V_o. Numerical values must be shown.

     (B) Calculate the **r.m.s.** value of the output voltage V_o.

   (iii) The resistance of resistor R_f is gradually increased from 3 kΩ to 8 kΩ. Describe what happens to the output voltage V_o during this time. 7

(9)

Marks

27. A ray of red light is directed at a glass prism of side 80 mm as shown in the diagram below.

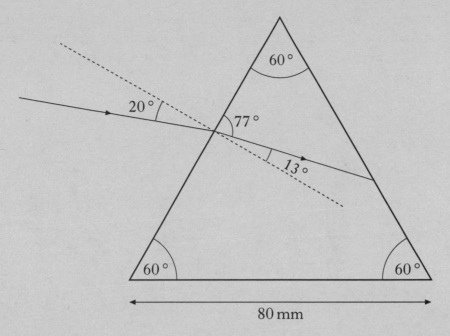

(a) Using information from this diagram, show that the refractive index of the glass for this red light is 1·52.   **1**

(b) What is meant by the term *critical angle*?   **1**

(c) Calculate the critical angle for the red light in the prism.   **2**

(d) Sketch a diagram showing the path of the ray of red light until after it leaves the prism. Mark on your diagram the values of all relevant angles.   **3**

(7)

[Turn over

Marks

28. An image intensifier is used to improve night vision. It does this by amplifying the light from an object.

Light incident on a photocathode causes the emission of photoelectrons. These electrons are accelerated by an electric field and strike a phosphorescent screen causing it to emit light. This emitted light is of a greater intensity than the light that was incident on the photocathode.

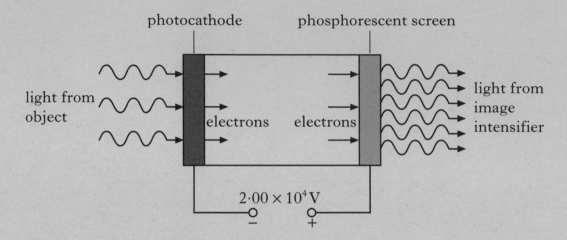

The voltage between the photocathode and the phosphorescent screen is $2 \cdot 00 \times 10^4$ V.

The minimum frequency of the incident light that allows photoemission to take place is $3 \cdot 33 \times 10^{14}$ Hz.

(a) What name is given to the minimum frequency of the light required for photoemission to take place?

1

(b) (i) Show that the work function of the photocathode material is $2 \cdot 21 \times 10^{-19}$ J.

(ii) Light of frequency $5 \cdot 66 \times 10^{14}$ Hz is incident on the photocathode. Calculate the maximum kinetic energy of an electron emitted from the photocathode.

(iii) Calculate the kinetic energy gained by an electron as it is accelerated from the photocathode to the phosphorescent screen.

6

(7)

Marks

29. (*a*) A sample of pure semiconducting material is doped by adding impurity atoms.

How does this addition affect the resistance of the semiconducting material? *Reduced*

1

(*b*) The circuit below shows a p-n junction diode used as a light emitting diode (LED).

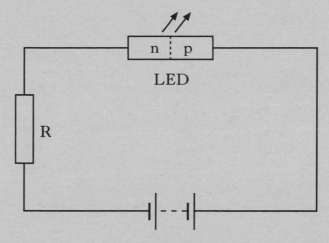

(i) Explain in terms of the charge carriers how the LED emits light.

(ii) Monochromatic light from the LED is incident on a grating as shown.

The spacing between lines in the grating is 5.0×10^{-6} m.

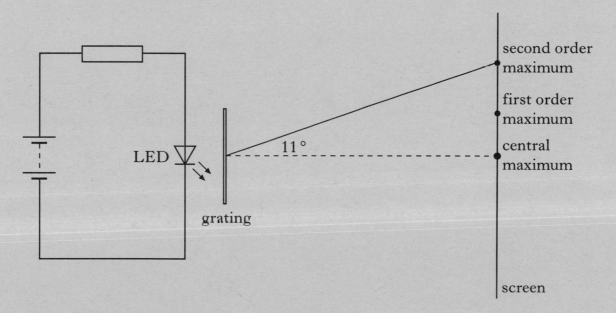

What is the wavelength of the light emitted by the LED?

4

(5)

[Turn over for Question 30 on *Page twenty-four*

Marks

30. (a) Torbernite is a mineral which contains uranium.

The activity of $1 \cdot 0\,kg$ of pure torbernite is $5 \cdot 9 \times 10^6$ decays per second.

A sample of material of mass $0 \cdot 6\,kg$ contains 40% torbernite. The remaining 60% of the material is not radioactive.

What is the activity of the sample in becquerels? **2**

(b) The table below gives the quality factor for some types of radiation.

Type of radiation	Quality factor
Gamma rays	1
Fast neutrons	10
Alpha particles	20

Exposure to $150\,\mu Gy$ of alpha particles for 6 hours gives the same dose equivalent rate as exposure for 8 hours to $400\,\mu Gy$ of one of the other radiations in the table above.

Identify this radiation.

You must justify your answer by calculation. **3**

(5)

[END OF QUESTION PAPER]

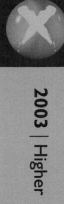

2003 | Higher

[BLANK]

X069/301

| NATIONAL QUALIFICATIONS 2003 | MONDAY, 19 MAY 1.00 PM – 3.30 PM | PHYSICS HIGHER |

Read Carefully

1 All questions should be attempted.

Section A (questions 1 to 20)

2 Check that the answer sheet is for Physics Higher (Section A).

3 Answer the questions numbered 1 to 20 on the answer sheet provided.

4 Fill in the details required on the answer sheet.

5 Rough working, if required, should be done only on this question paper, or on the first two pages of the answer book provided—**not** on the answer sheet.

6 For each of the questions 1 to 20 there is only **one** correct answer and each is worth 1 mark.

7 Instructions as to how to record your answers to questions 1–20 are given on page three.

Section B (questions 21 to 29)

8 Answer questions numbered 21 to 29 in the answer book provided.

9 Fill in the details on the front of the answer book.

10 Enter the question number clearly in the margin of the answer book beside each of your answers to questions 21 to 29.

11 Care should be taken to give an appropriate number of significant figures in the final answers to calculations.

SCOTTISH QUALIFICATIONS AUTHORITY

©

DATA SHEET
COMMON PHYSICAL QUANTITIES

Quantity	Symbol	Value	Quantity	Symbol	Value
Speed of light in vacuum	c	3.00×10^{8} m s^{-1}	Mass of electron	m_e	9.11×10^{-31} kg
Magnitude of the charge on an electron	e	1.60×10^{-19} C	Mass of neutron	m_n	1.675×10^{-27} kg
Gravitational acceleration on Earth	g	9.8 m s^{-2}	Mass of proton	m_p	1.673×10^{-27} kg
Planck's constant	h	6.63×10^{-34} J s			

REFRACTIVE INDICES
The refractive indices refer to sodium light of wavelength 589 nm and to substances at a temperature of 273 K.

Substance	Refractive index	Substance	Refractive index
Diamond	2·42	Water	1·33
Crown glass	1·50	Air	1·00

SPECTRAL LINES

Element	Wavelength/nm	Colour	Element	Wavelength/nm	Colour
Hydrogen	656	Red	Cadmium	644	Red
	486	Blue-green		509	Green
	434	Blue-violet		480	Blue
	410	Violet			
	397	Ultraviolet			
	389	Ultraviolet			
Sodium	589	Yellow			

Lasers		
Element	Wavelength/nm	Colour
Carbon dioxide	9550 } 10590	Infrared
Helium-neon	633	Red

PROPERTIES OF SELECTED MATERIALS

Substance	Density/ kg m^{-3}	Melting Point/ K	Boiling Point/ K
Aluminium	2.70×10^{3}	933	2623
Copper	8.96×10^{3}	1357	2853
Ice	9.20×10^{2}	273	. . . .
Sea Water	1.02×10^{3}	264	377
Water	1.00×10^{3}	273	373
Air	1·29	. . . .	. . . .
Hydrogen	9.0×10^{-2}	14	20

The gas densities refer to a temperature of 273 K and a pressure of 1.01×10^{5} Pa.

SECTION A

For questions 1 to 20 in this section of the paper, an answer is recorded on the answer sheet by indicating the choice A, B, C, D or E by a stroke made in ink in the appropriate box of the answer sheet—see the example below.

EXAMPLE

The energy unit measured by the electricity meter in your home is the

    A ampere

    B kilowatt-hour

    C watt

    D coulomb

    E volt.

The correct answer to the question is B—kilowatt-hour. Record your answer by drawing a heavy vertical line joining the two dots in the appropriate box on your answer sheet in the column of boxes headed B. The entry on your answer sheet would now look like this:

If after you have recorded your answer you decide that you have made an error and wish to make a change, you should cancel the original answer and put a vertical stroke in the box you now consider to be correct. Thus, if you want to change an answer D to an answer B, your answer sheet would look like this:

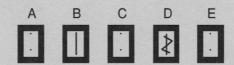

If you want to change back to an answer which has already been scored out, you should enter a tick (✓) to the RIGHT of the box of your choice, thus:

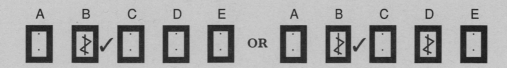

          Page three   **[Turn over**

SECTION A

Answer questions 1–20 on the answer sheet.

1. Which of the following are **both** vectors?

   A   Speed and weight

   B   Kinetic energy and potential energy

   C   Mass and momentum

   D   Weight and momentum

   E   Force and speed

2. A vehicle is travelling in a straight line.

   Graphs of velocity and acceleration against time are shown below.

   Which pair of graphs could represent the motion of the vehicle?

   A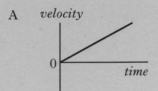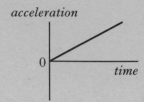

   B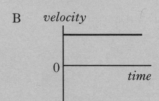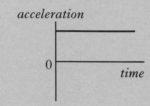

   C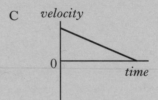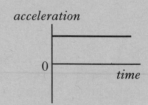

   D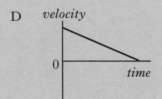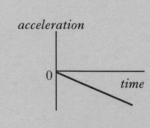

   E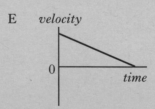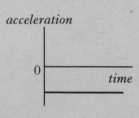

3. A block of mass 4·0 kg and a block of mass 6·0 kg are linked by a spring balance of negligible mass.

The blocks are placed on a frictionless horizontal surface. A force of 18·0 N is applied to the 6·0 kg block as shown.

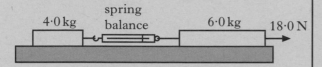

What is the reading on the spring balance?

A   7·2 N

B   9·0 N

C   10·8 N

D   18·0 N

E   40·0 N

4. A car of mass 1000 kg is travelling at a speed of 40 m s^{-1} along a straight road. The brakes are applied and the car decelerates to 10 m s^{-1}.

How much kinetic energy is lost by the car?

A   15 kJ

B   50 kJ

C   450 kJ

D   750 kJ

E   800 kJ

5. A car is designed with a "crumple-zone" so that the front of the car collapses during impact.

The purpose of the crumple-zone is to

A   decrease the driver's change in momentum per second

B   increase the driver's change in momentum per second

C   decrease the driver's final velocity

D   increase the driver's total change in momentum

E   decrease the driver's total change in momentum.

6. A fixed mass of gas condenses at atmospheric pressure to form a liquid.

Which row in the table shows the approximate increase in density and the approximate decrease in spacing between molecules?

	Approximate increase in density	Approximate decrease in spacing between molecules
A	10 times	2 times
B	100 times	10 times
C	1000 times	10 times
D	1 000 000 times	100 times
E	1 000 000 times	1000 times

[Turn over

7. A rigid metal cylinder stores compressed gas. Gas is gradually released from the cylinder. The temperature of the gas remains constant.

Which set of graphs shows how the pressure, the volume and the mass of the gas **in the cylinder** change with time?

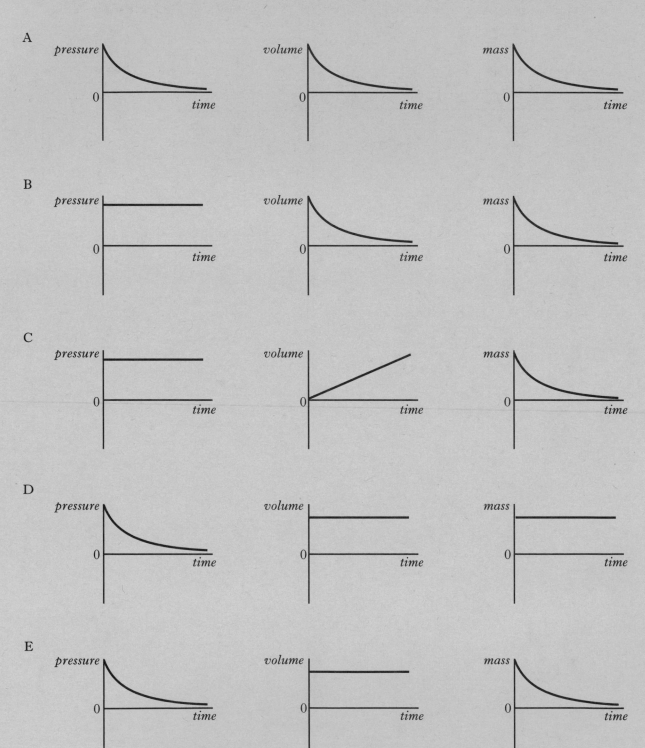

8. Two parallel metal plates, R and S, are connected to a 2·0 V d.c. supply as shown.

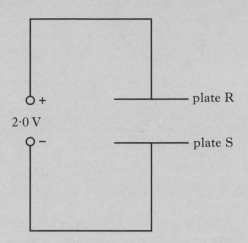

An electron is moved from plate R to plate S.

The gain in electrical potential energy of the electron is

A $8·0 \times 10^{-20}$ J

B $1·6 \times 10^{-19}$ J

C $3·2 \times 10^{-19}$ J

D $6·4 \times 10^{-19}$ J

E $1·3 \times 10^{-19}$ J.

9. In the following circuit, the battery has an e.m.f. of 8·0 V and an internal resistance of 0·20 Ω.

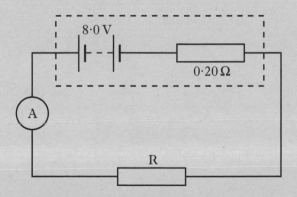

The reading on the ammeter is 4·0 A.

The resistance of R is

A   0·5 Ω

B   1·8 Ω

C   2·0 Ω

D   2·2 Ω

E   6·4 Ω.

10. In the following circuit, the supply has negligible internal resistance.

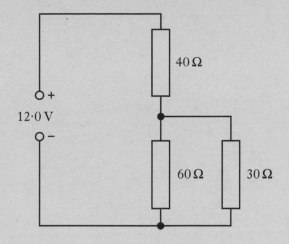

The p.d. across the 30 Ω resistor is

A   8·0 V

B   7·2 V

C   6·0 V

D   4·8 V

E   4·0 V.

11. A student sets up the following circuit.

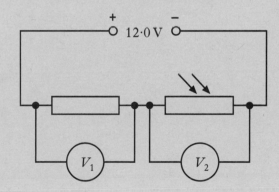

The intensity of light incident on the LDR is reduced.

Which row in the table shows the effect on the voltmeter readings V_1 and V_2?

	V_1	V_2
A	increases	increases
B	decreases	decreases
C	increases	decreases
D	decreases	increases
E	no change	increases

12. A student writes the following statements about a capacitor.

   I The current in a circuit containing a capacitor decreases when the supply frequency increases.

  II A capacitor can store charge.

III A capacitor can block d.c.

Which of these is/are correct?

A   I only

B   II only

C   III only

D   I and II only

E   II and III only

13. A farad is a

A   volt per ampere

B   volt per ohm

C   coulomb per volt

D   coulomb per second

E   joule per coulomb.

14. A $10\,\mu\text{F}$ capacitor is connected to a 50 V supply. The maximum charge stored by the capacitor is

A $2\!\cdot\!0 \times 10^{-7}\,\text{C}$

B $5\!\cdot\!0 \times 10^{-4}\,\text{C}$

C $5\!\cdot\!0\,\text{C}$

D $5\!\cdot\!0 \times 10^{2}\,\text{C}$

E $5\!\cdot\!0 \times 10^{6}\,\text{C}.$

15. In the following passage three words have been replaced by the letters **X**, **Y** and **Z**.

"*Monochromatic light is incident on a grating and the resulting interference pattern is viewed on a screen. The distance between neighbouring areas of constructive interference on the screen:*

*is ....**X**........... when the screen is moved further away from the grating;*

*is ....**Y**.......... when light of a greater wavelength is used;*

*is ....**Z**........... when the distance between the slits is increased.*"

Which row of the table shows the missing words?

	X	Y	Z
A	increased	increased	increased
B	increased	increased	decreased
C	decreased	decreased	increased
D	decreased	decreased	decreased
E	increased	decreased	decreased

16. An engineer creates an experimental window using sheets of transparent plastics **P**, **Q** and **R**.

A ray of light directed at the window follows the path shown.

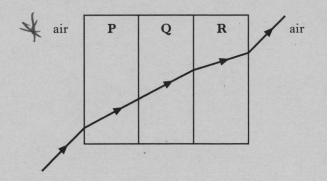

Which row in the table gives possible values for the refractive indices of the three plastics?

	P	*Q*	*R*
A	1·5	1·9	2·3
B	1·5	1·5	2·3
C	2·3	2·3	1·5
D	2·3	1·9	1·5
E	1·5	1·5	1·2

17. A unit for the intensity of light is

A $J m^{-1}$

B $J m^{-2}$

C $J s^{-1} m^{-1}$

D $J s^{-1} m^{-2}$

E $J s^{-2} m^{-2}$.

18. When light of frequency f is shone on to a certain metal, photoelectrons are ejected with a maximum velocity v and kinetic energy E_k.

When light of the same frequency and twice the intensity is shone on the same surface then

 I twice as many electrons are ejected per second

 II the speed of the fastest electrons is now $2v$

 III the kinetic energy of the fastest electrons is now $2E_k$.

Which of the statements above is/are correct?

A I only

B II only

C III only

D I and II only

E II and III only

19. A student writes the following statements about n-type semiconductor material.

 I Most charge carriers are negative.

 II The n-type material has a negative charge.

 III Impurity atoms in the material have 5 outer electrons.

Which of these statements is/are true?

A I only

B II only

C III only

D I and II only

E I and III only

20. Which of the following statements describes nuclear fission?

A A nucleus of large mass number splits into two nuclei, releasing several neutrons.

B A nucleus of large mass number splits into two nuclei, releasing several electrons.

C A nucleus of large mass number splits into two nuclei, releasing several protons.

D Two nuclei combine to form one nucleus, releasing several electrons.

E Two nuclei combine to form one nucleus, releasing several neutrons.

[SECTION B begins on *Page eleven*]

SECTION B

Write your answers to questions 21 to 29 in the answer book.

Marks

21. A golfer on an elevated tee hits a golf ball with an initial velocity of $35 \cdot 0\,\mathrm{m\,s^{-1}}$ at an angle of $40°$ to the horizontal.

    The ball travels through the air and hits the ground at point R.

    Point R is 12 m below the height of the tee, as shown.

    diagram not to scale

    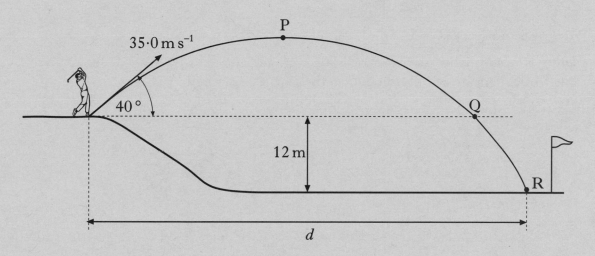

    The effects of air resistance can be ignored.

    (a) Calculate:

    (i) the horizontal component of the initial velocity of the ball;

    (ii) the vertical component of the initial velocity of the ball;

    (iii) the time taken for the ball to reach its maximum height at point P.   4

    (b) From its maximum height at point P, the ball falls to point Q, which is at the same height as the tee.

    It then takes a further $0 \cdot 48$ s to travel from Q until it hits the ground at R.

    Calculate the total horizontal distance d travelled by the ball.   3

    (7)

    [Turn over

Marks

22. Two ice skaters are initially skating together, each with a velocity of $2 \cdot 2 \, \text{m s}^{-1}$ to the right as shown.

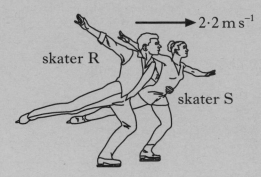

The mass of skater R is $54 \, \text{kg}$. The mass of skater S is $38 \, \text{kg}$.

Skater R now pushes skater S with an average force of $130 \, \text{N}$ for a short time. This force is in the same direction as their original velocity.

As a result, the velocity of skater S increases to $4 \cdot 6 \, \text{m s}^{-1}$ to the right.

(a) Calculate the magnitude of the change in momentum of skater S.　　　　2

(b) How long does skater R exert the force on skater S?　　　　2

(c) Calculate the velocity of skater R immediately after pushing skater S.　　　　2

(d) Is this interaction between the skaters elastic?

You must justify your answer by calculation.　　　　3

(9)

Marks

23. A tank of water rests on a smooth horizontal surface.

    (*a*) A student takes measurements of the pressure at various depths below the surface of the water, using the apparatus shown.

    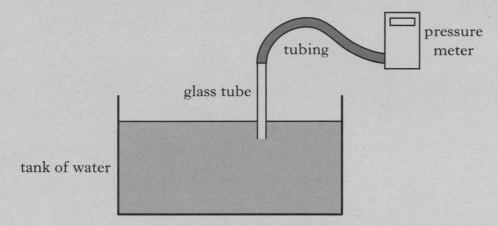

    The pressure meter is set to zero before the glass tube is lowered into the water.

    (i) Sketch a graph to show how the pressure due to the water varies with depth below the surface of the water.

    (ii) Calculate the pressure due to the water at a depth of 0·25 m below its surface.

    (iii) As the glass tube is lowered further into the tank, the student notices that some water rises inside the glass tube. Explain why this happens.

    4

    (*b*) The mass of water in the tank is $2·7 \times 10^3$ kg. The tank has a mass of 300 kg and a flat rectangular base. The dimensions of the tank are shown in the diagram below.

    Atmospheric pressure is $1·01 \times 10^5$ Pa.

    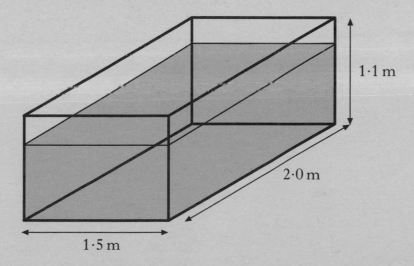

    Calculate the total pressure exerted by the base of the tank on the surface on which it rests.

    3

    (7)

Marks

24. A technician designs the following apparatus to investigate the pressure of a gas at different temperatures.

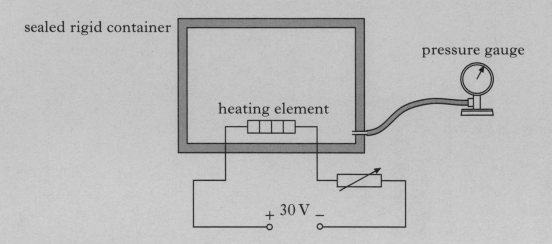

sealed rigid container

pressure gauge

heating element

+ 30 V –

The heating element is used to raise the temperature of the gas.

(a) Initially the gas is at a pressure of $1·56 \times 10^5$ Pa and a temperature of 27 °C. The temperature of the gas is then raised by 50 °C.

Calculate the new pressure of the gas in the container.   2

(b) The power supply shown above has an e.m.f. of 30 V and negligible internal resistance. The resistance of the heating element is $0·50 \, \Omega$ and the resistance of the variable resistor is set to $1·50 \, \Omega$.

(i) Calculate the power output from the heating element.

(ii) How would your answer to part (b)(i) be affected if the internal resistance of the power supply was **not** negligible? You must justify your answer.   4

(6)

Marks

25. (*a*) A signal generator is connected to an oscilloscope. The output of the signal generator is set to a peak voltage of 15 V.

The following diagram shows the trace obtained, the Y-gain and the timebase controls on the oscilloscope. The scale for the Y-gain has been omitted.

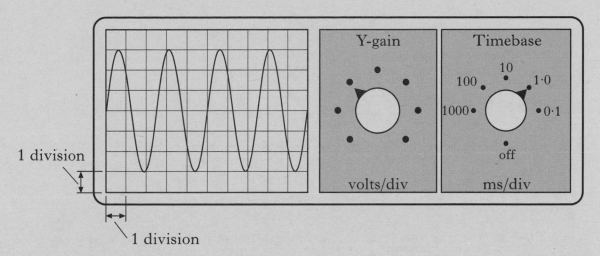

Calculate:

  (i)   the Y-gain setting of the oscilloscope;

  (ii) the frequency of the signal in hertz. **3**

(*b*) The signal generator is now connected in the circuit shown below.

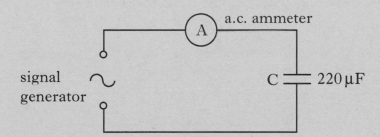

The signal generator is adjusted to give a peak output voltage of 12 V at a frequency of 300 Hz. The internal resistance of the signal generator and the resistance of the a.c. ammeter are negligible.

  (i)   Calculate the r.m.s. value of the output voltage from the signal generator.

  (ii) Calculate the **maximum** energy stored by the capacitor during one cycle of the supply voltage.

  (iii) The frequency of the signal generator is gradually increased.

        What happens to the reading on the ammeter?

  (iv) When a capacitor is connected to a d.c. supply, the current quickly falls to zero. Explain why the current does **not** fall to zero in the circuit above. **6**

(9)

[Turn over

Marks

26. A washing machine is filled with water, emptied and refilled several times during a wash cycle. A water level detector is used to ensure the water does not overflow.

One design of water level detector uses a specially shaped glass prism, as shown below.

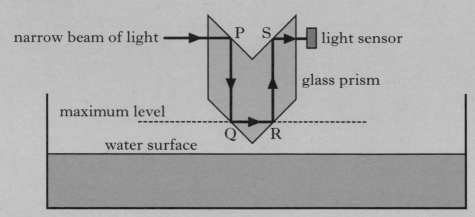

When the water in the machine is below the maximum level indicated in the diagram, the light sensor is illuminated by the narrow beam of light.

(a) The light sensor consists of an LDR connected in a Wheatstone bridge circuit with values of resistance as shown.

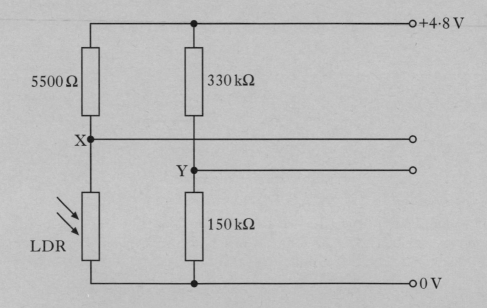

When the water in the machine is at the maximum level, the bridge is balanced.

Calculate the resistance of the LDR when the bridge is balanced.

2

Marks

26.   (continued)

(b)   Points X and Y of the Wheatstone bridge are connected to the inputs of an op-amp circuit as shown.

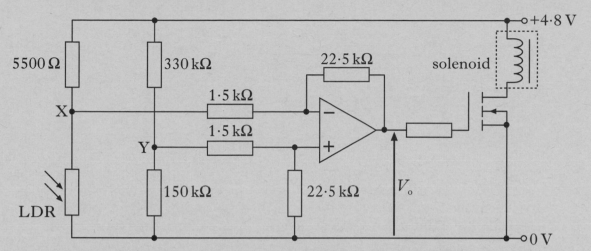

The potential at Y is 1·50 V. When the washing machine is filling with water, the narrow beam of light illuminates the LDR, the bridge is unbalanced and the potential at X is 1·28 V.

(i)   Name the component in the circuit which has the following symbol.

(ii)   Calculate the output voltage V_o of the op-amp when the LDR is illuminated.

(iii)   When there is a current in the solenoid, it holds a valve open and water flows into the washing machine.

When the water reaches the maximum level, the valve closes.

Explain how the circuit causes the valve to close when the water reaches the maximum level.

5

(c)   When the water is at the maximum level, the narrow beam of light no longer illuminates the light sensor, because light leaves the prism at Q.

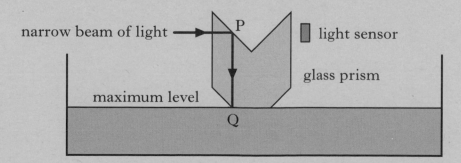

Explain why the light leaves the prism at Q.

1

(8)

Marks

27. (*a*) Electrons which orbit the nucleus of an atom can be considered as occupying discrete energy levels.

The following diagram shows some of the energy levels for a particular atom.

$$E_4 \text{———————} -1 \cdot 4 \times 10^{-19} \text{J}$$
$$E_3 \text{———————} -2 \cdot 4 \times 10^{-19} \text{J}$$

$$E_2 \text{———————} -5 \cdot 6 \times 10^{-19} \text{J}$$

$$E_1 \text{———————} -21 \cdot 8 \times 10^{-19} \text{J}$$

   (i) The transition between which two of these energy levels produces radiation with the longest wavelength? You must justify your answer.

   (ii) Calculate the frequency of the photon produced when an electron falls from E_3 to E_2.

5

(*b*) A laser produces light of frequency $4 \cdot 74 \times 10^{14}$ Hz in air.

A ray of light from this laser is directed into a block of glass as shown below.

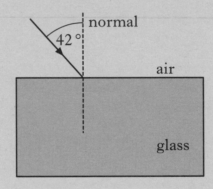

The refractive index of the glass for this light is $1 \cdot 60$.

   (i) What is the value of the frequency of the light in the block of glass?

   (ii) Calculate the wavelength of the light in the glass.

4

(9)

Marks

28. (*a*) An experiment with microwaves is set up as shown below.

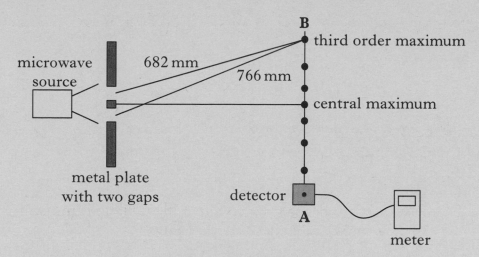

(i) As the detector is moved from **A** to **B**, the reading on the meter increases and decreases several times.

Explain, in terms of waves, how the pattern of maxima and minima is produced.

(ii) The measurements of the distance from each gap to a third order maximum are shown. Calculate the wavelength of the microwaves.   **3**

(*b*) A microphone is placed inside the cockpit of a jet aircraft.

The microphone is connected to the input terminals of the op-amp circuit shown below.

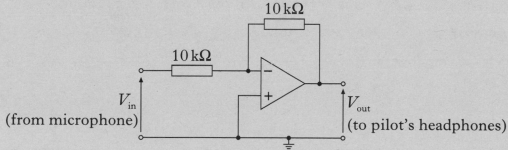

A noise in the cockpit produces the following signal from the microphone.

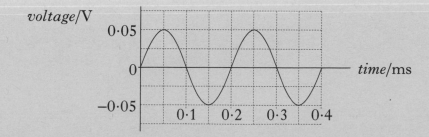

(i) Sketch a graph of the corresponding output voltage V_{out} against time.

Values are required on both axes.

(ii) The output from the op-amp is connected to the pilot's headphones.

Explain why the sound produced by the headphones **reduces** the noise level heard by the pilot.   **4**

Marks

29. A technician is studying samples of radioactive substances.

(a) The following statement describes a nuclear decay in one of the samples used by the technician.

$$^{238}_{92}U \rightarrow {}^{234}_{90}Th + {}^{4}_{2}He$$

    (i) What type of particle is emitted during this decay?

    (ii) In this sample $7 \cdot 2 \times 10^5$ nuclei decay in two minutes.

        Calculate the average activity of the sample during this time.   **3**

(b) The technician now studies the absorption of the radiation emitted from a different sample using the apparatus shown below.

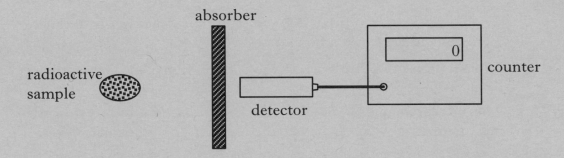

Different thicknesses of the absorber are placed in turn between the sample and the detector. For each thickness, the technician makes **repeated** measurements to obtain an average corrected count rate.

These results are then used to produce the following graph.

Marks

29. **(b)** **(continued)**

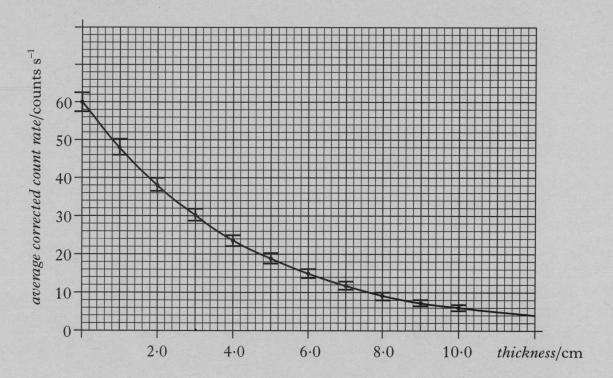

(i) Use the graph to calculate the half-value thickness of the absorber material for this radiation.

(ii) The technician has plotted each value of the average corrected count rate as a point with a vertical "bar" as shown.

Suggest a reason for this.　　　　　　　　　　　　　　　　　　**3**

(c) The technician receives a total dose equivalent of $6.4 \times 10^{-5}\,$Sv from these two sources.

The quality factor of the radiation used in part *(a)* is 20.

The absorbed dose received by the technician from the source used in part *(b)* is $1.2 \times 10^{-5}\,$Gy. The quality factor of this radiation is 1.

Calculate the absorbed dose received by the technician from the source used in part *(a)*.　　　　　　　　　　　　　　　　**2**

　　　　　　　　　　　　　　　　　　　　　　　　　　(8)

[END OF QUESTION PAPER]

[BLANK PAGE]

2004 | Higher

[BLANK]

X069/301

NATIONAL
QUALIFICATIONS
2004

FRIDAY, 28 MAY
1.00 PM – 3.30 PM

PHYSICS
HIGHER

Read Carefully

1   All questions should be attempted.

Section A (questions 1 to 20)

2   Check that the answer sheet is for Physics Higher (Section A).

3   Answer the questions numbered 1 to 20 on the answer sheet provided.

4   Fill in the details required on the answer sheet.

5   Rough working, if required, should be done only on this question paper, or on the first two pages of the answer book provided—**not** on the answer sheet.

6   For each of the questions 1 to 20 there is only **one** correct answer and each is worth 1 mark.

7   Instructions as to how to record your answers to questions 1–20 are given on page three.

Section B (questions 21 to 30)

8   Answer questions numbered 21 to 30 in the answer book provided.

9   Fill in the details on the front of the answer book.

10   Enter the question number clearly in the margin of the answer book beside each of your answers to questions 21 to 30.

11   Care should be taken to give an appropriate number of significant figures in the final answers to calculations.

Scottish
Qualifications
Authority

DATA SHEET
COMMON PHYSICAL QUANTITIES

Quantity	Symbol	Value	Quantity	Symbol	Value
Speed of light in vacuum	c	3.00×10^8 m s^{-1}	Mass of electron	m_e	9.11×10^{-31} kg
Magnitude of the charge on an electron	e	1.60×10^{-19} C	Mass of neutron	m_n	1.675×10^{-27} kg
Gravitational acceleration on Earth	g	9.8 m s^{-2}	Mass of proton	m_p	1.673×10^{-27} kg
Planck's constant	h	6.63×10^{-34} J s			

REFRACTIVE INDICES
The refractive indices refer to sodium light of wavelength 589 nm and to substances at a temperature of 273 K.

Substance	Refractive index	Substance	Refractive index
Diamond	2·42	Water	1·33
Crown glass	1·50	Air	1·00

SPECTRAL LINES

Element	Wavelength/nm	Colour	Element	Wavelength/nm	Colour
Hydrogen	656	Red	Cadmium	644	Red
	486	Blue-green		509	Green
	434	Blue-violet		480	Blue
	410	Violet			
	397	Ultraviolet			
	389	Ultraviolet			

Lasers		
Element	Wavelength/nm	Colour
Carbon dioxide	9550 } 10590 }	Infrared
Helium-neon	633	Red

Element	Wavelength/nm	Colour
Sodium	589	Yellow

PROPERTIES OF SELECTED MATERIALS

Substance	Density/ kg m^{-3}	Melting Point/ K	Boiling Point/ K
Aluminium	2.70×10^3	933	2623
Copper	8.96×10^3	1357	2853
Ice	9.20×10^2	273	. . . .
Sea Water	1.02×10^3	264	377
Water	1.00×10^3	273	373
Air	1.29	. . . .	. . . .
Hydrogen	9.0×10^{-2}	14	20

The gas densities refer to a temperature of 273 K and a pressure of 1.01×10^5 Pa.

SECTION A

For questions 1 to 20 in this section of the paper, an answer is recorded on the answer sheet by indicating the choice A, B, C, D or E by a stroke made in ink in the appropriate box of the answer sheet—see the example below.

EXAMPLE

The energy unit measured by the electricity meter in your home is the

    A ampere

    B kilowatt-hour

    C watt

    D coulomb

    E volt.

The correct answer to the question is B—kilowatt-hour. Record your answer by drawing a heavy vertical line joining the two dots in the appropriate box on your answer sheet in the column of boxes headed B. The entry on your answer sheet would now look like this:

If after you have recorded your answer you decide that you have made an error and wish to make a change, you should cancel the original answer and put a vertical stroke in the box you now consider to be correct. Thus, if you want to change an answer D to an answer B, your answer sheet would look like this:

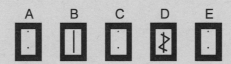

If you want to change back to an answer which has already been scored out, you should enter a tick (✓) to the RIGHT of the box of your choice, thus:

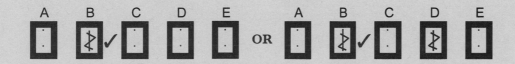

SECTION A

Answer questions 1–20 on the answer sheet.

1. A box is pulled along a level bench by a rope held at a constant angle of 40° to the horizontal as shown.

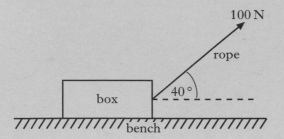

A constant force of 100 N is applied to the rope.

The box moves a distance of 10 m along the bench.

The work done on the box by the rope is

A   100 J

B   643 J

C   766 J

D   839 J

E   1000 J.

2. A stuntman on a motorcycle jumps a river which is 5·1 m wide. He lands on the edge of the far bank, which is 2·0 m lower than the bank from which he takes off.

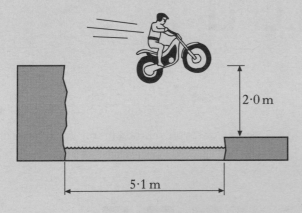

His minimum horizontal speed at take off is

A   2·0 m s^{-1}

B   3·2 m s^{-1}

C   5·5 m s^{-1}

D   8·0 m s^{-1}

E   9·8 m s^{-1}.

3. A vehicle of mass 0·1 kg is moving to the right along a horizontal friction-free air track. A vehicle of mass 0·2 kg is moving to the left on the same track.

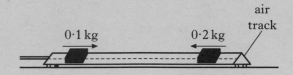

The vehicles collide and stick together.

Which of the following quantities is/are conserved in this collision?

 I   The total momentum

 II   The kinetic energy

 III The total energy

A   I only

B   II only

C   I and II only

D   I and III only

E   II and III only

4. A cannon of mass 1200 kg fires a cannonball of mass 15 kg at a velocity of 60 m s^{-1} East.

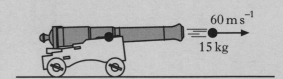

Assuming the force of friction is negligible, the velocity of the cannon just after firing is

A   0 m s^{-1}

B   0·75 m s^{-1} East

C   0·75 m s^{-1} West

D   6·0 m s^{-1} East

E   6·0 m s^{-1} West.

5. Car X is designed with a "crumple-zone" so that the front of the car collapses during impact as shown.

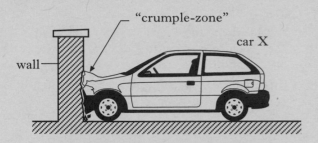

A similar car, Y, of equal mass is built without a crumple-zone. In a safety test both cars are driven at the same speed into identical walls.

Which of the following statements is/are true during the collisions?

  I The average force on car X is smaller than that on car Y.

 II The time taken for car X to come to rest is greater than that for car Y.

III The change in momentum of car X is smaller than that of car Y.

  A I only

  B I and II only

  C I and III only

  D II and III only

  E I, II and III

6. A golf ball, initially at rest, is hit by a club. The graph of the force of the club on the ball against time is shown.

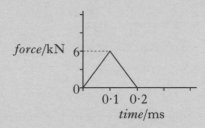

A different type of golf ball of the same size and mass is now hit with the same club. This ball moves off with the same velocity as the first ball.

Which graph shows the force of the club on the second ball against time?

A

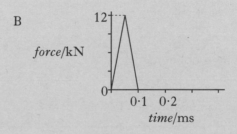

B

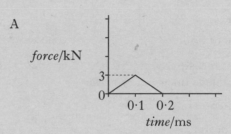

C

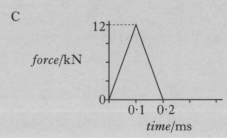

D

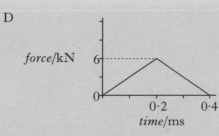

E

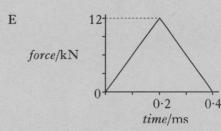

Page five

[Turn over

7. The density of steam at $100\,°C$ is less than the density of water at $100\,°C$. The explanation for this is that when water changes to steam its particles

   A move further apart

   B move with greater speed

   C have smaller mass

   D are no longer joined together

   E collide more often with each other.

8. In an experiment the following measurements and uncertainties are recorded.

   Temperature rise $= 10\,°C \pm 1\,°C$ ^10

   Heater current $= 5\cdot0\,A \pm 0\cdot2\,A$ ⁴

   Heater voltage $= 12\cdot0\,V \pm 0\cdot5\,V$ ⁴·²

   Time $= 100\,s \pm 2\,s$ ²

   Mass of liquid $= 1\cdot000\,kg \pm 0\cdot005\,kg$ ·⁵

   The measurement which has the largest percentage uncertainty is the

   A temperature rise

   B heater current

   C heater voltage

   D time

   E mass of liquid.

9. A balloon of volume of $6\cdot0\,m^3$ contains a fixed mass of gas at a temperature of $300\,K$ and a pressure of $2\cdot0\,kPa$. The gas is heated to $600\,K$ and the pressure reduced to $1\cdot0\,kPa$. The new volume of the gas is

   A $1\cdot5\,m^3$

   B $3\cdot0\,m^3$

   C $6\cdot0\,m^3$

   D $12\cdot0\,m^3$

   E $24\cdot0\,m^3$.

10. A student writes the following statements about electric fields.

    I There is a force on a charge in an electric field.

    II When an electric field is applied to a conductor, the free electric charges in the conductor move.

    III Work is done when a charge is moved in an electric field.

    Which of the above statements is/are correct?

    A I only

    B II only

    C I and II only

    D I and III only

    E I, II and III

11. A resistor and a capacitor are connected to identical a.c. supplies which provide constant voltage throughout their whole frequency range.

Which of the following pairs of graphs illustrates how the current varies with frequency in the two circuits shown?

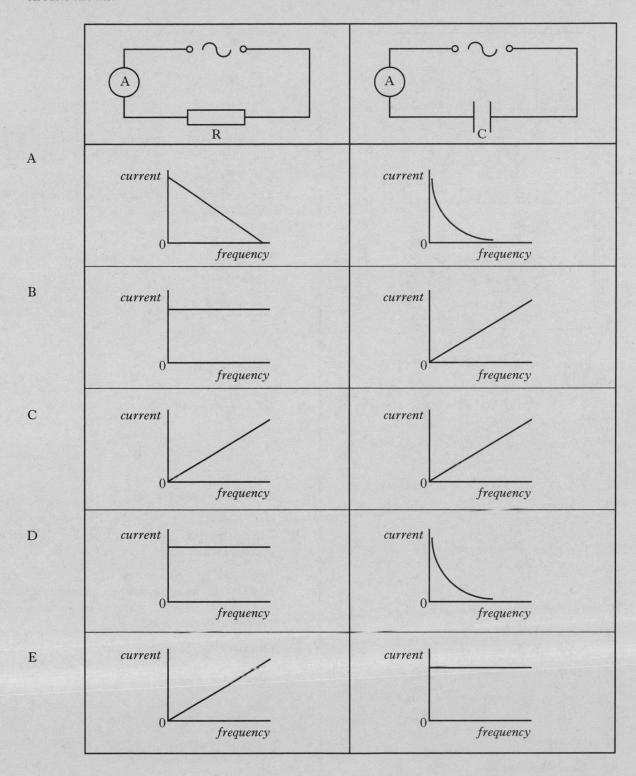

[Turn over

12. The output from a signal generator is connected to the input terminals of an oscilloscope. The trace observed on the oscilloscope screen, the Y-gain setting and the time-base setting are shown in the diagram.

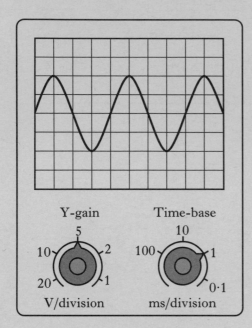

The frequency of the signal shown is calculated using the

A Y-gain setting and the vertical height of the trace

B Y-gain setting and the horizontal distance between the peaks of the trace

C Y-gain setting and time-base setting

D time-base setting and the vertical height of the trace

E time-base setting and the horizontal distance between the peaks of the trace.

13. Which of the following statements is/are true for an ideal op-amp?

  I It has infinite input resistance.

  II Both inputs are at the same potential.

  III The input current to the op-amp is zero.

A I only

B II only

C I and II only

D II and III only

E I, II and III

14. Two identical loudspeakers, L_1 and L_2, are operated at the same frequency and in phase with each other. An interference pattern is produced.

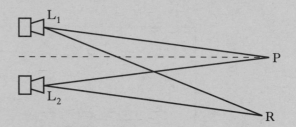

At position P, which is the same distance from both loudspeakers, there is a maximum intensity.

The next maximum intensity is at position R, where $L_1R = 5 \cdot 6\,\mathrm{m}$ and $L_2R = 5 \cdot 3\,\mathrm{m}$.

The speed of sound is $340\,\mathrm{m\,s^{-1}}$.

The frequency of the sound emitted by the loudspeakers is given by

A $\dfrac{5 \cdot 6 - 5 \cdot 3}{340}\,\mathrm{Hz}$

B $\dfrac{340}{5 \cdot 6 + 5 \cdot 3}\,\mathrm{Hz}$

C $\dfrac{340}{5 \cdot 6 - 5 \cdot 3}\,\mathrm{Hz}$

D $340 \times (5 \cdot 6 - 5 \cdot 3)\,\mathrm{Hz}$

E $340 \times (5 \cdot 6 + 5 \cdot 3)\,\mathrm{Hz}$.

15. Ultraviolet radiation is incident on a clean zinc plate. Photoelectrons are ejected.

The clean zinc plate is replaced by a different metal which has a lower work function. The same intensity of ultraviolet radiation is incident on this metal.

Compared to the zinc plate, which of the following statements is/are true for the new metal?

  I The maximum speed of the photoelectrons is greater.

  II The maximum kinetic energy of the photoelectrons is greater.

  III There are more photoelectrons ejected per second.

A I only

B II only

C III only

D I and II only

E I, II and III

16. An atom has the energy levels shown.

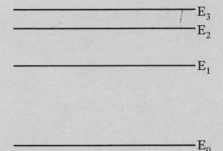

Electron transitions occur between all of these levels to produce emission lines in the spectrum of this atom.

How many emission lines are produced by transitions between these energy levels?

A   3

B   4

C   5

D   6

E   7

17. Materials are "doped" to produce n-type semiconductor material.

In n-type semiconductor material

A   the majority charge carriers are electrons

B   the majority charge carriers are neutrons

C   the majority charge carriers are protons

D   there are more protons than neutrons

E   there are more electrons than neutrons.

18. A student writes the following statements about the decay of radionuclides.

  I During alpha emission a particle consisting of 2 protons and 4 neutrons is emitted from a nucleus.

  II During beta emission a fast moving electron is emitted from a nucleus.

  III During gamma emission a high energy photon is emitted from a nucleus.

Which of these statements is/are true?

A   II only

B   I and II only

C   I and III only

D   II and III only

E   I, II and III

19. A radiation technician works 150 hours each month in an area exposed to radiation from a neutron beam. The quality factor for this radiation is 3. The technician receives an absorbed dose rate of $10\,\mu Gy\,h^{-1}$ from this radiation.

In a period of 5 months the total dose equivalent received by the technician is

A $2\cdot50 \times 10^{-2}\,Sv$

B $2\cdot25 \times 10^{-2}\,Sv$

C $1\cdot50 \times 10^{-2}\,Sv$

D $1\cdot00 \times 10^{-2}\,Sv$

E $0\cdot75 \times 10^{-2}\,Sv.$

20. A Geiger counter records a corrected count-rate of 1000 counts per second when it is placed a distance of 400 mm from a radioactive source.

A sheet of metal is placed between the source and the counter. The half value thickness of the metal for radiation from the source is 20 mm.

The corrected count-rate is now 125 counts per second.

The thickness of the metal sheet is

A   25 mm

B   40 mm

C   60 mm

D   80 mm

E   160 mm.

[Turn over

SECTION B

Write your answers to questions 21 to 30 in the answer book.

Marks

21. (*a*) State the difference between speed and velocity.　　　　　　　**1**

(*b*) During a tall ships race, a ship called the Mir passes a marker buoy X and sails due West (270). It sails on this course for 30 minutes at a speed of $10.0\,km\,h^{-1}$, then changes course to 20° West of North (340). The Mir continues on this new course for $1\frac{1}{2}$ hours at a speed of $8.0\,km\,h^{-1}$ until it passes marker buoy Y.

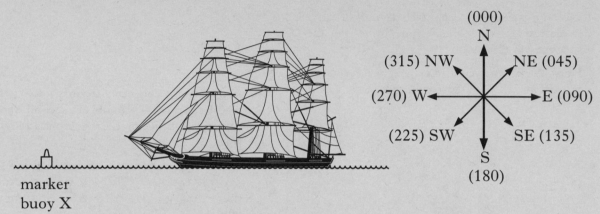

marker
buoy X

(i) Show that the Mir travels a total distance of 17 km between marker buoys X and Y.

(ii) By scale drawing or otherwise, find the displacement from marker buoy X to marker buoy Y.

(iii) Calculate the average velocity, in $km\,h^{-1}$, of the Mir between marker buoys X and Y.　　　　　　　**6**

(*c*) A second ship, the Leeuvin, passes marker buoy X 15 minutes after the Mir and sails directly for marker buoy Y at a speed of $7.5\,km\,h^{-1}$.

Show by calculation which ship first passes marker buoy Y.　　　　　　　**2**

(9)

Marks

22. A train of mass 7.5×10^5 kg is travelling at $60\,\text{m s}^{-1}$ along a straight horizontal track.

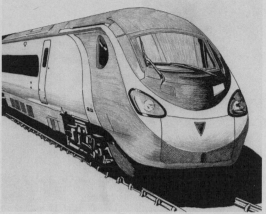

The brakes are applied and the train decelerates uniformly to rest in a time of 40 s.

(a) (i) Calculate the distance the train travels between the brakes being applied and the train coming to rest.

    (ii) Calculate the force required to bring the train to rest in this time.　**4**

(b) Part of the train's braking system consists of an electrical circuit as shown in the diagram.

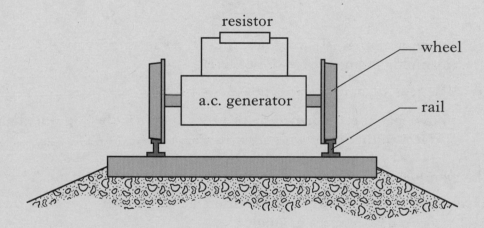

While the train is braking, the wheels drive an a.c. generator which changes kinetic energy into electrical energy. This electrical energy is changed into heat in a resistor. The r.m.s. current in the resistor is 2.5×10^3 A and the resistor produces 8.5 MJ of heat each second.

Calculate the peak voltage across the resistor.　**3**

(7)

[Turn over

Marks

23. A crane barge is used to place part of an oil well, called a manifold, on the seabed.

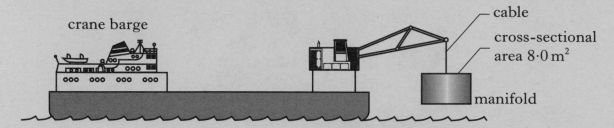

crane barge

cable

cross-sectional area $8 \cdot 0 \, \text{m}^2$

manifold

The manifold is a cylinder of uniform cross-sectional area $8 \cdot 0 \, \text{m}^2$ and mass $5 \cdot 0 \times 10^4 \, \text{kg}$. The mass of the cable may be ignored.

(*a*) Calculate the tension in the cable when the manifold is held stationary above the surface of the water.

1

(*b*) The manifold is lowered into the water and then held stationary just below the surface as shown.

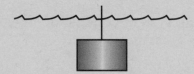

   (i) Draw a sketch showing all the forces acting vertically on the manifold. Name each of these forces.

  (ii) The tension in the cable is now $2 \cdot 5 \times 10^5 \, \text{N}$.

Show that the difference in pressure between the top and bottom surfaces of the manifold is $3 \cdot 0 \times 10^4 \, \text{Pa}$.

4

(*c*) The manifold is now lowered to a greater depth.

What effect does this have on the difference in pressure between the top and bottom surfaces of the manifold?

You must justify your answer.

2

(7)

Marks

24. A student sets up the circuit shown.

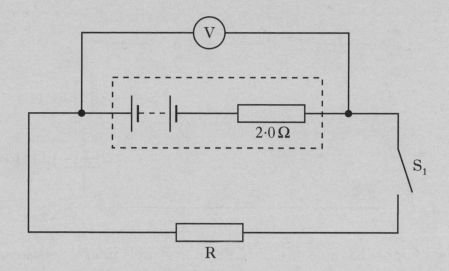

The internal resistance of the battery is $2 \cdot 0 \, \Omega$.

With S_1 open, the student notes that the reading on the voltmeter is $9 \cdot 0 \, V$.

The student closes S_1 and notes that the reading on the voltmeter is now $7 \cdot 8 \, V$.

(*a*) (i) Calculate the resistance of resistor R.

(ii) Explain why the reading on the voltmeter decreases when S_1 is closed.

4

(*b*) The student adds a $30 \, \Omega$ resistor and a switch S_2 to the circuit as shown.

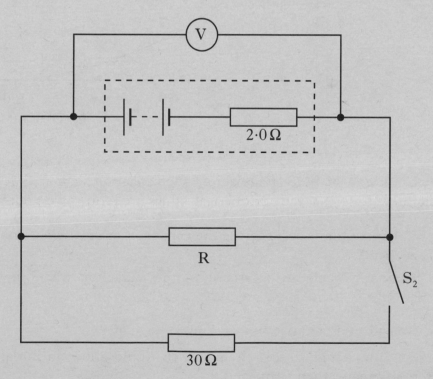

The student now closes S_2.

Explain what happens to the reading on the voltmeter.

2

(6)

Marks

25. In an experiment, the circuit shown is used to investigate the charging of a capacitor.

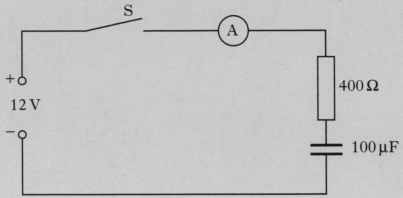

The power supply has an e.m.f. of 12 V and negligible internal resistance. The capacitor is initially uncharged.

Switch S is closed and the current measured during charging. The graph of charging current against time is shown in figure 1.

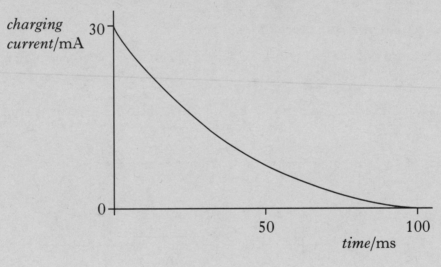

figure 1

(a) Sketch a graph of the voltage across the capacitor against time until the capacitor is fully charged. Numerical values are required on both axes.   **2**

(b) (i) Calculate the voltage across the capacitor when the charging current is 20 mA.

   (ii) How much energy is stored in the capacitor when the charging current is 20 mA?   **4**

(c) The capacitor has a maximum working voltage of 12 V.

   Suggest **one** change to this circuit which would allow an initial charging current of greater than 30 mA.   **1**

Marks

25. **(continued)**

(*d*) The $100\,\mu\text{F}$ capacitor is now replaced by an uncharged capacitor of unknown capacitance and the experiment is repeated. The graph of charging current against time for this capacitor is shown in figure 2.

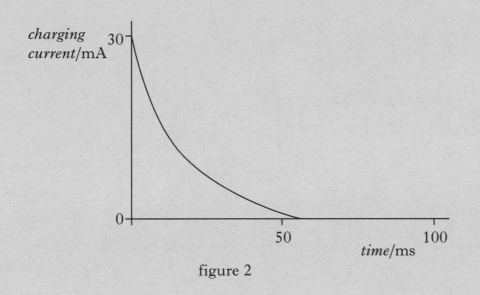

figure 2

By comparing figure 2 with figure 1, determine whether the capacitance of this capacitor is greater than, equal to or less than $100\,\mu\text{F}$.

You must justify your answer.

2

(9)

[Turn over

Marks

26. The circuit shown is designed for an alarm system.

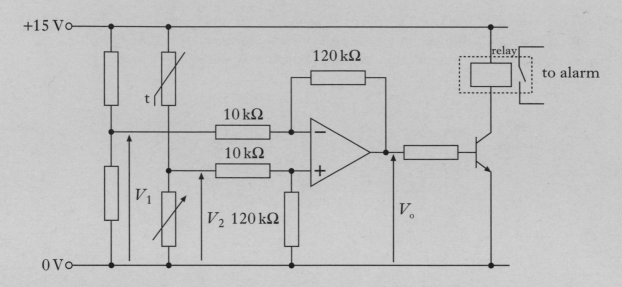

Voltage V_1 is 7·50 V.

When the temperature increases, the resistance of the thermistor decreases.

(a) At a temperature of 35 °C, voltage V_2 is 7·52 V.

Calculate the output voltage V_o at this temperature. **2**

(b) When the temperature rises, V_o increases and the alarm switches on.

Explain how the circuit operates to switch on the alarm. **2**

(c) The alarm is on when V_o is greater than or equal to 0·72 V.

The graph of the temperature against voltage V_2 is shown.

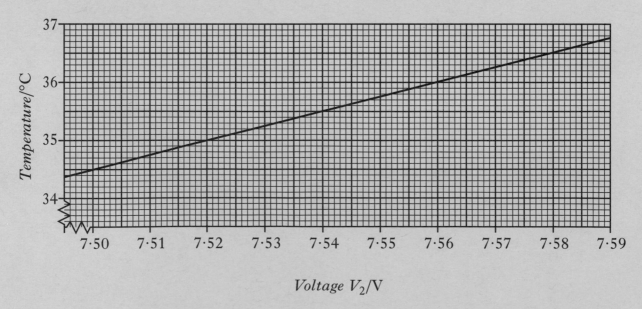

Using information from the graph, determine the minimum temperature at which the alarm switches on. **2**

(6)

Marks

27. A decorative lamp has a transparent liquid in the space above a bulb. Light from the bulb passes through rotating coloured filters giving red or blue light in the liquid.

(a) A ray of red light is incident on the liquid surface as shown.

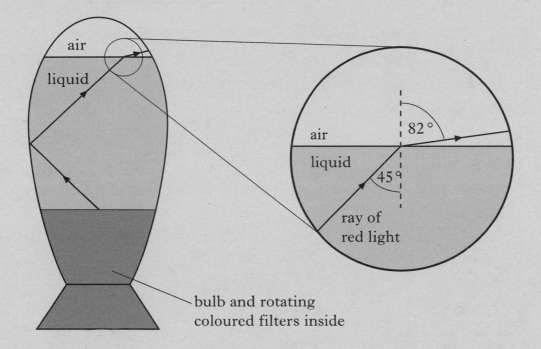

bulb and rotating
coloured filters inside

(i) Calculate the refractive index of the liquid for the red light.

(ii) A ray of blue light is incident on the liquid surface at the same angle as the ray of red light.

The refractive index of the liquid for blue light is greater than that for red light. Is the angle of refraction greater than, equal to or less than 82° for the blue light?

You must explain your answer. **3**

(b) A similar lamp contains a liquid which has a refractive index of 1·44 for red light. A ray of red light in the liquid is incident on the surface at an angle of 45° as before.

Sketch a diagram to show the path of this ray after it is incident on the liquid surface.

Mark on your diagram the values of all appropriate angles.

All relevant calculations must be shown. **3**

(6)

[Turn over

Marks

28. The term LASER is short for "Light Amplification by the Stimulated Emission of Radiation".

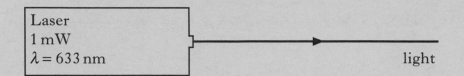

(a) (i) Describe what is meant by *Stimulated Emission*.

(ii) Explain how amplification is produced in a laser. **3**

(b) In an experiment, laser light of wavelength 633 nm is incident on a grating.

A series of bright spots are seen on a screen placed some distance from the grating. The distance between these spots and the central spot is shown.

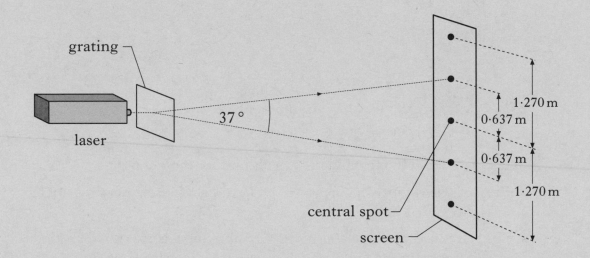

Calculate the number of lines per metre on the grating. **3**

(c) The laser is replaced with another laser and the experiment repeated. With this laser the bright spots are closer together.

How does the wavelength of the light from this laser compare with that from the original laser?

You must justify your answer. **2**

(8)

Marks

29. An LED consists of a p-n junction as shown.

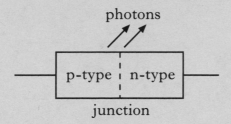

(a) Copy the diagram and add a battery so that the p-n junction is forward-biased.   1

(b) Using the terms *electrons, holes* and *photons*, explain how light is produced at the p-n junction of the LED.   1

(c) The LED emits photons, of energy $3 \cdot 68 \times 10^{-19}$ J.

   (i) Calculate the wavelength of a photon of light from this LED.

   (ii) Calculate the minimum potential difference across the p-n junction whcn it emits photons.   4

                                                                                                                                                                         (6)

[Turn over for Question 30 on *Page twenty*

Marks

30. A ship is powered by a nuclear reactor.

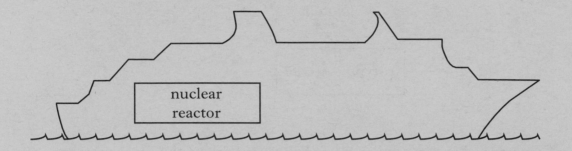

One reaction that takes place in the core of the nuclear reactor is represented by the statement below.

$$^{235}_{92}\text{U} + ^{1}_{0}\text{n} \rightarrow ^{140}_{58}\text{Ce} + ^{94}_{40}\text{Zr} + 2^{1}_{0}\text{n} + 6^{0}_{-1}\text{e}$$

(a) The symbol for the Uranium nucleus is $^{235}_{92}\text{U}$.

What information about the nucleus is provided by the following numbers?

(i) 92

(ii) 235   **2**

(b) Describe how neutrons produced during the reaction can cause further nuclear reactions.   **1**

(c) The masses of particles involved in the reaction are shown in the table.

Particles	Mass/kg
$^{235}_{92}\text{U}$	$390 \cdot 173 \times 10^{-27}$
$^{140}_{58}\text{Ce}$	$232 \cdot 242 \times 10^{-27}$
$^{94}_{40}\text{Zr}$	$155 \cdot 884 \times 10^{-27}$
$^{1}_{0}\text{n}$	$1 \cdot 675 \times 10^{-27}$
$^{0}_{-1}\text{e}$	negligible

Calculate the energy released in the reaction.   **3**

     (6)

[END OF QUESTION PAPER]

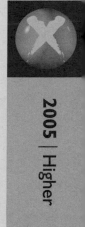

2005 | Higher

[BLANK PAGE]

X069/301

NATIONAL
QUALIFICATIONS
2005

TUESDAY, 24 MAY
1.00 PM – 3.30 PM

PHYSICS
HIGHER

Read Carefully

1 All questions should be attempted.

Section A (questions 1 to 20)

2 Check that the answer sheet is for Physics Higher (Section A).

3 Check that the answer sheet you have been given has **your name**, **date of birth**, **SCN** (Scottish Candidate Number) and **Centre Name** printed on it.

Do not change any of these details.

4 If any of this information is wrong, tell the Invigilator immediately.

5 If this information is correct, **print** your name and seat number in the boxes provided.

6 Use **black** or **blue ink** for your answers. **Do not use red ink**.

7 There is **only one correct** answer to each question.

8 Any rough working should be done on the question paper or the rough working sheet, **not** on your answer sheet.

9 At the end of the exam, put the **answer sheet for Section A inside the front cover of your answer book**.

10 Instructions as to how to record your answers to questions 1–20 are given on page three.

Section B (questions 21 to 30)

11 Answer questions numbered 21 to 30 in the answer book provided.

12 Fill in the details on the front of the answer book.

13 Enter the question number clearly in the margin of the answer book beside each of your answers to questions 21 to 30.

14 Care should be taken to give an appropriate number of significant figures in the final answers to calculations.

SCOTTISH
QUALIFICATIONS
AUTHORITY

DATA SHEET
COMMON PHYSICAL QUANTITIES

Quantity	Symbol	Value	Quantity	Symbol	Value
Speed of light in vacuum	c	$3.00 \times 10^{8} \, \text{m s}^{-1}$	Mass of electron	m_e	$9.11 \times 10^{-31} \, \text{kg}$
Magnitude of the charge on an electron	e	$1.60 \times 10^{-19} \, \text{C}$	Mass of neutron	m_n	$1.675 \times 10^{-27} \, \text{kg}$
Gravitational acceleration on Earth	g	$9.8 \, \text{m s}^{-2}$	Mass of proton	m_p	$1.673 \times 10^{-27} \, \text{kg}$
Planck's constant	h	$6.63 \times 10^{-34} \, \text{J s}$			

REFRACTIVE INDICES
The refractive indices refer to sodium light of wavelength 589 nm and to substances at a temperature of 273 K.

Substance	Refractive index	Substance	Refractive index
Diamond	2·42	Water	1·33
Crown glass	1·50	Air	1·00

SPECTRAL LINES

Element	Wavelength/nm	Colour	Element	Wavelength/nm	Colour
Hydrogen	656	Red	Cadmium	644	Red
	486	Blue-green		509	Green
	434	Blue-violet		480	Blue
	410	Violet	*Lasers*		
	397	Ultraviolet	Element	Wavelength/nm	Colour
	389	Ultraviolet	Carbon dioxide	9550 } 10590 }	Infrared
Sodium	589	Yellow	Helium-neon	633	Red

PROPERTIES OF SELECTED MATERIALS

Substance	Density/ kg m^{-3}	Melting Point/ K	Boiling Point/ K
Aluminium	2.70×10^{3}	933	2623
Copper	8.96×10^{3}	1357	2853
Ice	9.20×10^{2}	273	. . . .
Sea Water	1.02×10^{3}	264	377
Water	1.00×10^{3}	273	373
Air	1·29	. . . .	. . . .
Hydrogen	9.0×10^{-2}	14	20

The gas densities refer to a temperature of 273 K and a pressure of 1.01×10^{5} Pa.

SECTION A

For questions 1 to 20 in this section of the paper the answer to each question is either A, B, C, D or E. Decide what your answer is, then put a horizontal line in the space provided—see the example below.

EXAMPLE

The energy unit measured by the electricity meter in your home is the

    A   ampere

    B   kilowatt-hour

    C   watt

    D   coulomb

    E   volt.

The correct answer is **B**—kilowatt-hour. The answer **B** has been clearly marked with a horizontal line (see below).

Changing an answer

If you decide to change your answer, cancel your first answer by putting a cross through it (see below) and fill in the answer you want. The answer below has been changed to **B**.

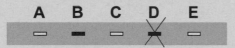

If you then decide to change back to an answer you have already scored out, put a tick (✓) to the **right** of the answer you want, as shown below:

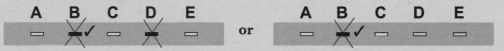

[Turn over

SECTION A

Answer questions 1–20 on the answer sheet.

1. A car travels from X to Y and then from Y to Z as shown.

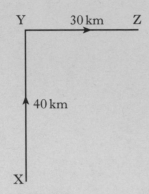

It takes one hour to travel from X to Y. It also takes one hour to travel from Y to Z.

Which row in the following table shows the magnitudes of the displacement, average speed and average velocity for the complete journey?

	Displacement (km)	Average speed (km h^{-1})	Average velocity (km h^{-1})
A	50	35	25
B	70	35	25
C	50	35	35
D	70	70	50
E	50	70	25

2. An object has a constant acceleration of $3\,m\,s^{-2}$. This means that the

A distance travelled by the object increases by 3 metres every second

B displacement of the object increases by 3 metres every second

C speed of the object is $3\,m\,s^{-1}$ every second

D velocity of the object is $3\,m\,s^{-1}$ every second

E velocity of the object increases by $3\,m\,s^{-1}$ every second.

3. A car of mass $1200\,kg$ pulls a horsebox of mass $700\,kg$ along a straight, horizontal road. They have an acceleration of $2\cdot0\,m\,s^{-2}$.

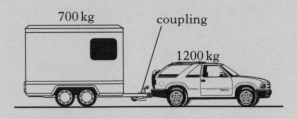

Assuming that the frictional forces are negligible, the tension in the coupling between the car and the horsebox is

A 500 N

B 700 N

C 1400 N

D 2400 N

E 3800 N.

4. A mass of $2\,kg$ slides along a frictionless surface at $10\,m\,s^{-1}$ and collides with a stationary mass of $10\,kg$.

before impact

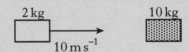

After the collision, the $2\,kg$ mass rebounds at $5\,m\,s^{-1}$ and the $10\,kg$ mass moves off at $3\,m\,s^{-1}$.

after impact

Which row in the following table is correct?

	Momentum of system	Kinetic energy of system	Type of collision
A	conserved	conserved	elastic
B	conserved	not conserved	inelastic
C	conserved	not conserved	elastic
D	not conserved	not conserved	inelastic
E	not conserved	not conserved	elastic

5. A golfer hits a ball of mass 5.0×10^{-2} kg with a golf club. The ball leaves the tee with a velocity of $80\,\text{m s}^{-1}$. The club is in contact with the ball for a time of $0.10\,\text{s}$.

The average force exerted by the club on the ball is

A $6.25 \times 10^{-4}\,\text{N}$

B $0.025\,\text{N}$

C $0.4\,\text{N}$

D $4\,\text{N}$

E $40\,\text{N}$.

6. A solid at a temperature of $-20\,°\text{C}$ is heated until it becomes a liquid at $70\,°\text{C}$.

The temperature change in kelvin is

A $50\,\text{K}$

B $90\,\text{K}$

C $343\,\text{K}$

D $363\,\text{K}$

E $596\,\text{K}$.

7. One volt is

A   one coulomb per joule

B   one joule coulomb

C   one joule per coulomb

D   one joule per second

E   one coulomb per second.

8. A potential difference of $5000\,\text{V}$ is applied between two metal plates. The plates are $0.10\,\text{m}$ apart. A charge of $+2.0\,\text{mC}$ is released from rest at the positively charged plate as shown.

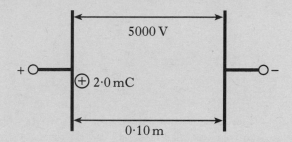

The kinetic energy of the charge just before it hits the negative plate is

A $4.0 \times 10^{-7}\,\text{J}$

B $2.0 \times 10^{-4}\,\text{J}$

C $5.0\,\text{J}$

D $10\,\text{J}$

E $500\,\text{J}$.

9. An a.c. signal is displayed on an oscilloscope screen. The Y-gain and time-base controls are set as shown.

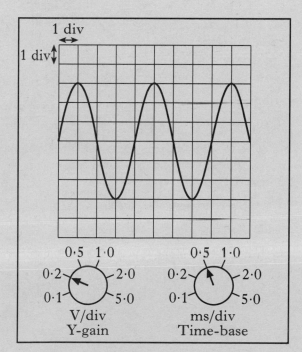

The frequency of the signal is

A $0.50\,\text{Hz}$

B $1.25\,\text{Hz}$

C $2.00\,\text{Hz}$

D $200\,\text{Hz}$

E $500\,\text{Hz}$.

10. A capacitor is connected to a circuit as shown.

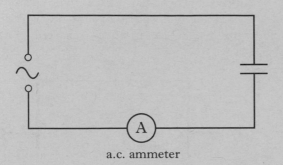

a.c. ammeter

The alternating supply has a constant peak voltage but its frequency can be varied.

The frequency is steadily increased from 50 Hz to 5000 Hz. The reading on the a.c. ammeter

A   remains constant

B   decreases steadily

C   increases steadily

D   increases then decreases

E   decreases then increases.

11. An amplifier circuit is shown.

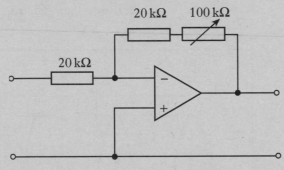

The variable resistor can be adjusted from zero to 100 kΩ. This allows the voltage gain to be altered over the range

A   zero to one

B   zero to five

C   zero to six

D   one to five

E   one to six.

12. A student sets up the following circuit.

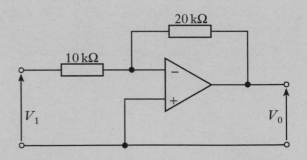

The graph below shows how the input voltage V_1 varies with time.

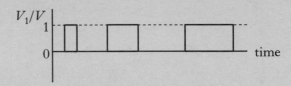

Which of the following graphs shows how the output voltage V_0 varies with time?

A
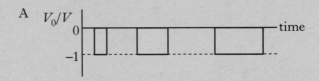

B
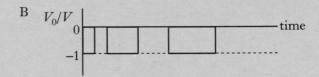

C
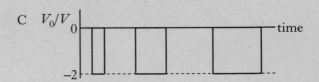

D
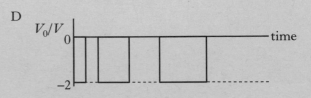

E
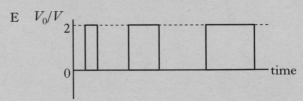

13. The apparatus used to investigate the relationship between light intensity I and distance d from a point source is shown.

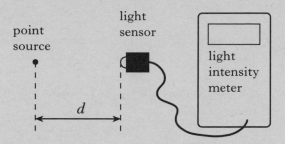

The experiment is carried out in a darkened room.

Which of the following expressions gives a constant value?

A $\quad I \times d$

B $\quad I \times d^2$

C $\quad \dfrac{I}{d}$

D $\quad \dfrac{I}{d^2}$

E $\quad I \times \sqrt{d}$

14. Microwaves of frequency $2 \cdot 0 \times 10^{10}\,\text{Hz}$ travel through air with a speed of $3 \cdot 0 \times 10^8\,\text{m s}^{-1}$. On entering a bath of oil, the speed reduces to $1 \cdot 5 \times 10^8\,\text{m s}^{-1}$.

The frequency of the microwaves in the oil is

A $\quad 1 \cdot 0 \times 10^{10}\,\text{Hz}$

B $\quad 2 \cdot 0 \times 10^{10}\,\text{Hz}$

C $\quad 4 \cdot 0 \times 10^{10}\,\text{Hz}$

D $\quad 3 \cdot 0 \times 10^{18}\,\text{Hz}$

E $\quad 6 \cdot 0 \times 10^{18}\,\text{Hz.}$

15. Which graph shows the relationship between frequency f and wavelength λ of photons of electromagnetic radiation?

A

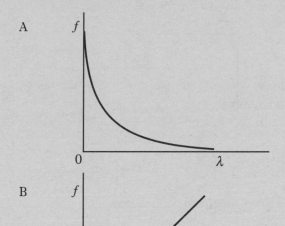

B

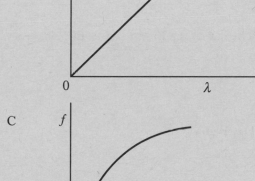

C

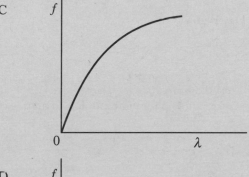

D

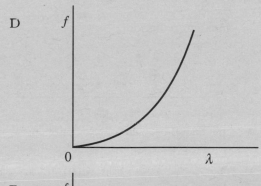

E

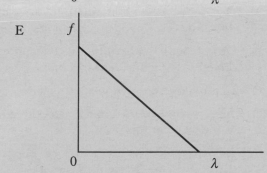

Page seven **[Turn over**

16. A liquid and a solid have the same refractive index.

What happens to the speed and the wavelength of light passing from the liquid into the solid?

	Speed	Wavelength
A	stays the same	stays the same
B	decreases	decreases
C	decreases	increases
D	increases	increases
E	increases	decreases

17. The intensity of light can be measured in

A   W

B $W\,m^{-1}$

C $W\,m$

D $W\,m^{-2}$

E $W\,m^{2}$.

18. The diagram below represents part of the process of stimulated emission in a laser.

electron   excited level state

incident radiation

ground level state

Which of the following statements best describes the emitted radiation?

A   Out of phase and emitted in the same direction as the incident radiation

B   Out of phase and emitted in the opposite direction to the incident radiation

C   In phase and emitted in all directions.

D   In phase and emitted in the same direction as the incident radiation

E   In phase and emitted in the opposite direction to the incident radiation

19. Part of a radioactive decay series is shown.

$$_{Q}^{P}\text{Bi} \xrightarrow[\text{decay}]{\beta} \ _{S}^{R}\text{Po} \xrightarrow[\text{decay}]{\alpha} \ _{82}^{208}\text{Pb}$$

A bismuth nucleus emits a beta particle and its product, a polonium nucleus, emits an alpha particle.

Which numbers are represented by P, Q, R and S?

	P	Q	R	S
A	212	85	212	84
B	212	83	212	84
C	212	85	208	83
D	210	83	208	81
E	210	85	210	84

20. The equation below represents a nuclear reaction.

$$_{92}^{235}\text{U} + _{0}^{1}\text{n} \longrightarrow \ _{36}^{92}\text{Kr} + _{56}^{141}\text{Ba} + _{0}^{1}\text{n} + _{0}^{1}\text{n} + _{0}^{1}\text{n}$$

It is an example of

A   nuclear fusion

B   alpha particle emission

C   beta particle emission

D   induced nuclear fission

E   spontaneous nuclear fission.

[SECTION B begins on *Page ten*]

[Turn over

Marks

SECTION B

Write your answers to questions 21 to 30 in the answer book.

21. (*a*) A student uses the apparatus shown to measure the average acceleration of a trolley travelling down a track.

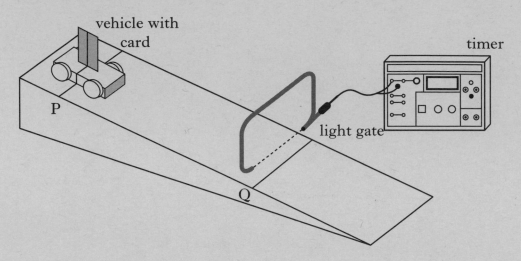

The line on the trolley is aligned with line P on the track.

The trolley is released from rest and allowed to run down the track.

The timer measures the time for the card to pass through the light gate.

This procedure is repeated a number of times and the results shown below.

   0·015 s   0·013 s   0·014 s   0·019 s   0·017 s   0·018 s

   (i) Calculate:

   (A) the mean time for the card to pass through the light gate;   **1**

   (B) the approximate absolute random uncertainty in this value.   **1**

   (ii) The length of the card is 0·020 m and the distance PQ is 0·60 m.

   Calculate the acceleration of the trolley (an uncertainty in this value is not required).   **3**

21. **(continued)**

(b) The light gate consists of a lamp shining onto a photodiode.

The photodiode forms part of the circuit shown.

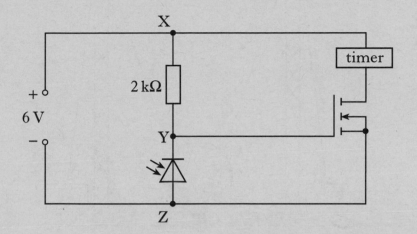

(i) In which mode is the photodiode operating? **1**

(ii) Explain why the timer only operates while the light beam is broken. **2**

(8)

[Turn over

Marks

22. A "giant catapult" is part of a fairground ride.

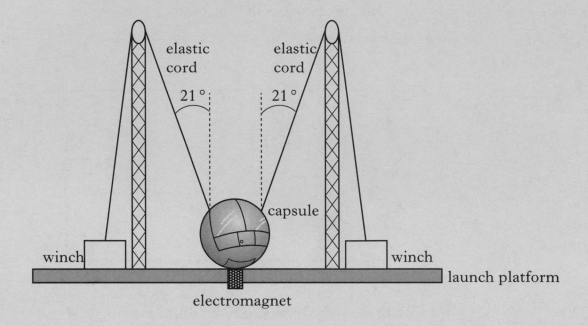

Two people are strapped into a capsule. The capsule and the occupants have a combined mass of 236 kg.

The capsule is held stationary by an electromagnet while the tension in the elastic cords is increased using the winches.

The mass of the elastic cords and the effects of air resistance can be ignored.

(a) When the tension in each cord reaches 4.5×10^3 N the electromagnet is switched off and the capsule and occupants are propelled vertically upwards.

   (i) Calculate the vertical component of the force exerted by **each** cord just before the capsule is released.   **1**

   (ii) Calculate the initial acceleration of the capsule.   **3**

   (iii) Explain why the acceleration of the capsule decreases as it rises.   **1**

(b) Throughout the ride the occupants remain upright in the capsule.

A short time after release the occupants feel no force between themselves and the seats.

Explain why this happens.   **1**

    (6)

Marks

23. A polystyrene float is held with its base 2·0 m below the surface of a swimming pool.

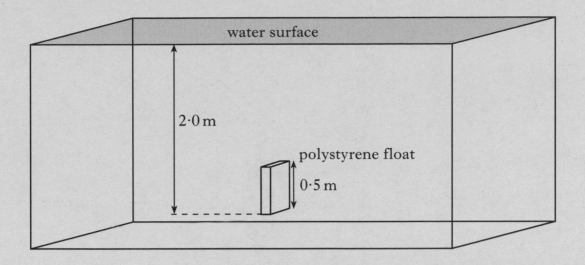

(a) The float has a mass of 12 g and its dimensions are 0·50 m × 0·30 m × 0·10 m.

Calculate the density of the float. **2**

(b) Explain why a buoyancy force acts on the float. **2**

(c) The float is released and accelerates towards the surface. Taking into account the resistance of the water, state what happens to the acceleration of the float as it approaches the surface. You must justify your answer. **2**

(d) Another float made of a more dense material with the same dimensions is now held at the same position in the pool.

The float is released as in part (c).

State how the initial acceleration of this float compares with the polystyrene float. You must justify your answer. **1**

(7)

[Turn over

Marks

24. The apparatus used to investigate the relationship between volume and temperature of a fixed mass of air is shown.

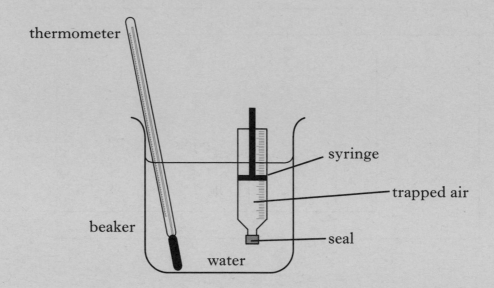

The volume of the trapped air is read from the scale on the syringe.

The temperature of the trapped air is altered by heating the water in the beaker. It is assumed that the temperature of the air in the syringe is the same as that of the surrounding water. The pressure of the trapped air is constant during the investigation.

(a) Readings of volume and temperature for the trapped air are shown.

Temperature/°C	25	50	75	100
Volume/ml	20·6	22·6	24·0	25·4

    (i) Using **all** the data, establish the relationship between temperature and volume for the trapped air.   **2**

    (ii) Calculate the volume of the trapped air when the temperature of the water is 65 °C.   **2**

    (iii) Use the kinetic model of gases to explain the change in volume as the temperature increases in this investigation.   **2**

Marks

24. (continued)

(b) An alternative to measuring the volume using the scale on the syringe, is to connect the piston of the syringe to a variable resistor.

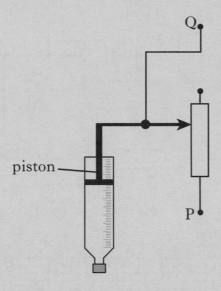

The variable resistor forms part of the circuit shown.

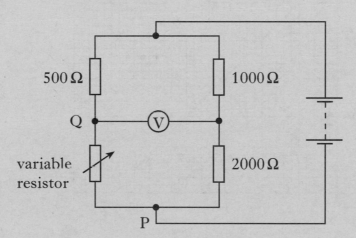

The reading on the voltmeter is 0 V when the temperature of the air in the syringe is 50 °C.

(i) Calculate the resistance of the variable resistor at this temperature.   2

(ii) The temperature of the gas in the syringe changes from just below to just above 50 °C. This causes the resistance of the variable resistor to change by a small amount.

Sketch a graph of the reading on the centre-zero voltmeter against the change in resistance of the variable resistor. Numerical values are not required on either axis.   1

(9)

[Turn over

Marks

25. A student sets up the following circuit to find the e.m.f. *E* and the internal resistance *r* of a battery.

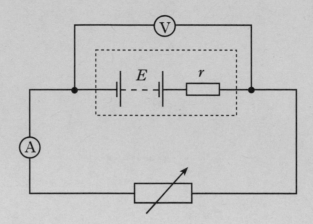

Readings from the voltmeter and ammeter are used to plot the following graph.

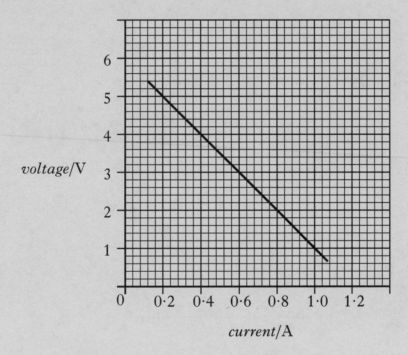

(*a*) What is meant by the term *e.m.f.*? 1

(*b*) (i) Use the graph to determine:

(A) the e.m.f.; 1

(B) the internal resistance of the battery. 2

(ii) Show that the variable resistor has a value of 15 Ω when the current is 0·30 A. 1

Marks

25. (continued)

(c) Without adjusting the variable resistor, a 30 Ω resistor is connected in parallel with it.

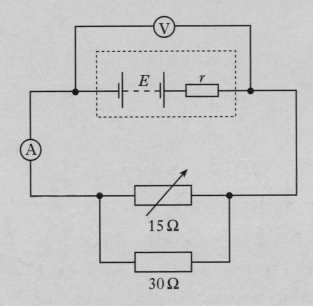

Calculate the new reading on the ammeter.

2

(7)

[Turn over

26. A student investigates the charging and discharging of a 2200 µF capacitor using the circuit shown.

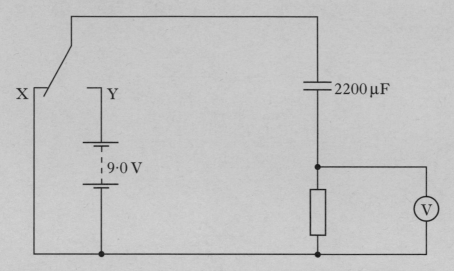

The 9·0 V battery has negligible internal resistance.

Initially the capacitor is uncharged and the switch is at position X.

The switch is then moved to position Y and the capacitor charges fully in 1·5 s.

(a) (i) Sketch a graph of the p.d. across the **resistor** against time while the capacitor charges. Appropriate numerical values are required on both axes. **2**

    (ii) The resistor is replaced with one of higher resistance.

       Explain how this affects the time taken to fully charge the capacitor. **1**

    (iii) At one instant during the charging of the capacitor the reading on the voltmeter is 4·0 V.

       Calculate the charge stored by the capacitor at this instant. **3**

(b) Using the same circuit in a later investigation the resistor has a resistance of 100 kΩ. The switch is in **position Y** and the capacitor is fully charged.

    (i) Calculate the maximum energy stored in the capacitor. **2**

    (ii) The switch is moved to position X. Calculate the maximum current in the resistor. **2**

(10)

27. A car is fitted with an alarm which sounds a buzzer when the outside temperature falls below 3 °C.

The sensor is a thermistor located under the mirror on the side of the car.

sensor

The thermistor forms part of the circuit shown.

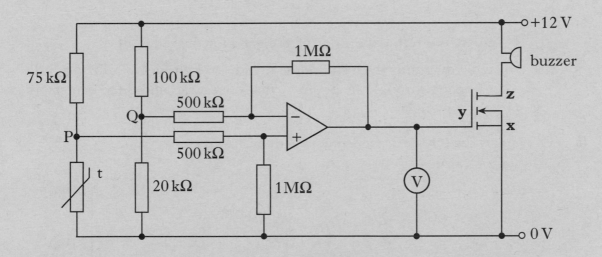

(a) What names are given to the terminals labelled **x**, **y** and **z** on the symbol for the MOSFET?

Clearly indicate which name goes with which letter.   **1**

(b) The buzzer sounds when the reading on the voltmeter is greater than or equal to +2·0 V.

   (i) Calculate the minimum potential difference required between points P and Q to sound the buzzer.   **2**

   (ii) Calculate the resistance of the thermistor when the reading on the voltmeter is +2·0 V.   **2**
                                                    (5)

[Turn over

Marks

28. A physics student investigates what happens when monochromatic light passes through a glass prism or a grating.

(*a*) The apparatus for the first experiment is shown below.

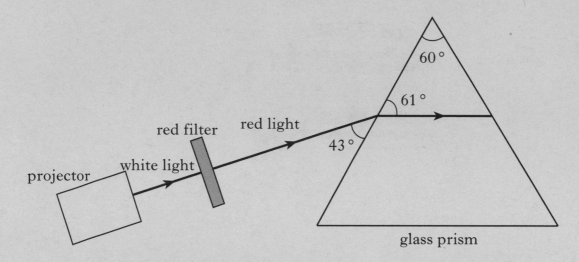

(i) Calculate the refractive index of the glass for the red light. **2**

(ii) Sketch a diagram which shows the ray of red light before, during and after passing through the prism. Mark on your diagram the values of all relevant angles. **2**

(*b*) The apparatus for the second experiment is shown below.

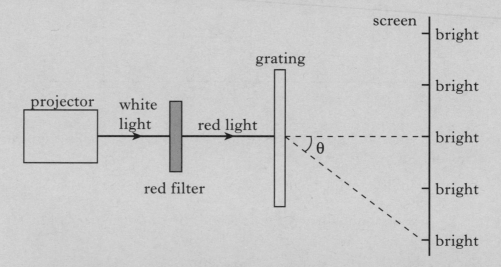

A pattern of bright and dark fringes is observed on the screen.

The grating has 300 lines per millimetre and the wavelength of the red light is 650 nm.

(i) Explain how the bright fringes are produced. **1**

(ii) Calculate the angle θ of the second order maximum. **2**

(iii) The red filter is replaced by a blue filter. Describe the effect of this change on the pattern observed.
Justify your answer. **1**

(8)

Marks

29. In 1902, P. Lenard set up an experiment similar to the one shown below.

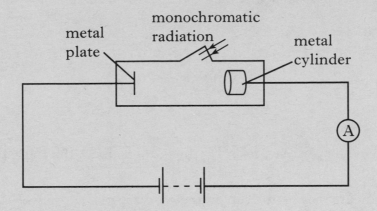

There is a constant potential difference between the metal plate and the metal cylinder.

Monochromatic radiation is directed onto the plate.

Photoelectrons produced at the plate are collected by the cylinder.

The frequency and the intensity of the radiation can be altered independently.

The frequency of the radiation is set at a value above the threshold frequency.

(*a*) The intensity of the radiation is slowly increased.

Sketch a graph of the current against intensity of radiation.   **1**

(*b*) The metal of the plate has a work function of $3 \cdot 11 \times 10^{-19}$ J. The wavelength of the radiation is 400 nm.

  (i)   Calculate the maximum kinetic energy of a photoelectron.   **3**

 (ii) The battery connections are now reversed.

Explain why there could still be a reading on the ammeter.   **1**

(5)

[Turn over

Marks

30. The nuclear industry must meet health and safety standards for workers. A worker has to handle radioactive materials behind a screen.

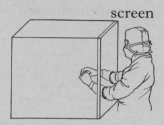

screen

(*a*) The screen must be sufficiently thick to reduce the radiation to an acceptable level.

Different thicknesses of the screen material are placed between the source and the Geiger-Muller tube.

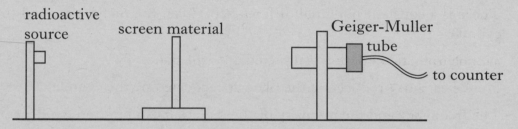

The graph shows corrected count rate plotted against thickness of material.

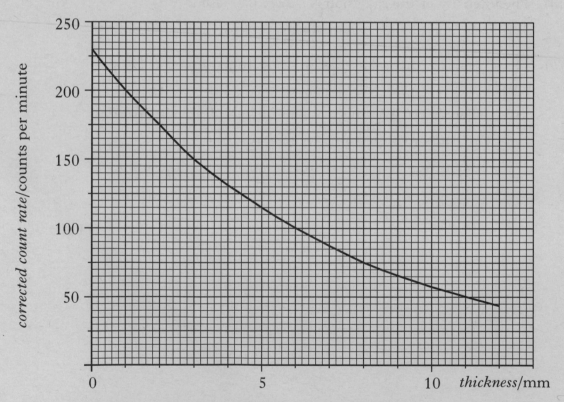

(i) Determine the half-value thickness of the material.   **1**

(ii) The dose equivalent rate in air a short distance from this source is $20\,\mu Sv\,h^{-1}$.

When a certain thickness of the material is placed in front of the source, the dose equivalent rate at the same distance falls to $2.5\,\mu Sv\,h^{-1}$.

Calculate the thickness of the material.   **2**

30. (continued)

(*b*) The recommended dose equivalent limit for exposure to the hands of a worker is 500 mSv per year.

On average the worker is exposed to 2·0 mGy of gamma radiation, 400 μGy of thermal neutrons and 80 μGy of fast neutrons each hour when working in this area.

The quality factors for these radiations are shown.

Radiation	Quality factor
gamma	1
thermal neutrons	3
fast neutrons	10

The recommended dose equivalent limit must not be exceeded.

Calculate the maximum number of working hours in one year permitted in this area.

2

(5)

[END OF QUESTION PAPER]

[BLANK PAGE]